CHAOS
FROM THEORY TO APPLICATIONS

CHAOS

FROM THEORY TO APPLICATIONS

ANASTASIOS A. TSONIS

University of Wisconsin at Milwaukee
Milwaukee, Wisconsin

PLENUM PRESS • NEW YORK AND LONDON

1992

Library of Congress Cataloging-in-Publication Data

Tsonis, Anastasios A.
   Chaos : from theory to applications / Anastasios A. Tsonis.
      p.   cm.
   Includes bibliographical references and index.
   ISBN 0-306-44171-3
   1. Chaotic behavior in systems.
Q172.5.C45T77   1992
003'.7--dc20   92-24411
                                                                    CIP

Cover illustration: A painting by the renowned Greek painter Takis Alexiou. It depicts the order within strange attractors and the randomness they generate.

ISBN 0-306-44171-3

© 1992 Plenum Press, New York
A Division of Plenum Publishing Corporation
233 Spring Street, New York, N.Y. 10013

Printed in the United States of America

To the smiles of my daughter Michelle

PREFACE

Based on chaos theory two very important points are clear: (1) random-looking aperiodic behavior may be the product of determinism, and (2) nonlinear problems should be treated as nonlinear problems and not as simplified linear problems.

The theoretical aspects of chaos have been presented in great detail in several excellent books published in the last five years or so. However, while the problems associated with applications of the theory—such as dimension and Lyapunov exponents estimation, chaos and nonlinear prediction, and noise reduction—have been discussed in workshops and articles, they have not been presented in book form.

This book has been prepared to fill this gap between theory and applications and to assist students and scientists wishing to apply ideas from the theory of nonlinear dynamical systems to problems from their areas of interest. The book is intended to be used as a text for an upper-level undergraduate or graduate-level course, as well as a reference source for researchers.

My philosophy behind writing this book was to keep it simple and informative without compromising accuracy. I have made an effort to present the concepts by using simple systems and step-by-step derivations. Anyone with an understanding of basic differential equations and matrix theory should follow the text without difficulty. The book was designed to be self-contained. When applicable, examples accompany the theory. The reader will notice, however, that in the later chapters specific examples become less frequent. This is purposely done in the hope that individuals will draw on their own ideas and research projects for examples.

I would like to thank Drs. P. Berge, M. Casdagli, J. Crutchfield, K. P. Georgakakos, L. Glass, A. Goldberger, C. Grebogi, E. J. Kostelich, M. D. Mundt, J. Nese, G. Nicolis, and J. Theiler for providing publication quality figures for use in the book. I would also like to thank Drs. M. Casdagli, D. Farmer, H. L. Swinney, and J. Theiler for providing me with reprints of their latest work. Thanks are also extended to all the scientists who gave me permission to reproduce figures from their papers. I also thank Donna Schenstron and my students Hong-Zhong Lu, Jianping Zhuang, John Roth, and Christine Young for producing several of the figures, and my friend and collaborator Dr. James Elsner for his ideas and discussions.

For me, writing this book was an experience. For the reader, I hope it will be a pleasure.

                                                        Anastasios A. Tsonis

Milwaukee, Wisconsin

CONTENTS

PART II: THEORY

PART III: APPLICATIONS

PART I

NOTES

CHAPTER 1

INTRODUCTION

Simplicity and regularity are associated with predictability. For example, because the orbit of the earth is simple and regular, we can always predict when astronomical winter will come. On the other hand, complexity and irregularity are almost synonymous with unpredictability.

Those who try to explain the world we live in always hope that in the realm of the complexity and irregularity observed in nature, simplicity would be found behind everything and, that, finally, unpredictable events would become predictable. That complexity and irregularity exist in nature is obvious. We need only look around us to realize that practically everything is random in appearance. Or is it? Clouds, like many other structures in nature, come in an infinite number of shapes. Every cloud is different, yet everybody will recognize a cloud. Clouds, though complex and irregular, must on the whole possess a uniqueness that distinguishes them from other structures in nature. The question remains: Is their irregularity completely random, or is there some order behind their irregularity?

Over the last decades physicists, mathematicians, astronomers, biologists, and scientists from many other disciplines have developed a new way of looking at complexity in nature. This way has been termed *chaos* theory. Chaos is mathematically defined as "randomness" generated by simple deterministic systems. This randomness is a result of the sensitivity of chaotic systems to the initial conditions. However, because the systems are deterministic, chaos implies some order. This interesting "mixture" of randomness and order allows us to take a different approach in studying processes that were thought to be completely random. Apparently, the founders of chaos theory had a very good sense of humor, since *chaos* is the Greek word for the complete absence of order.

The mathematical foundations of what is now called chaos were laid

3

down a long time ago by Poincaré in his work on bifurcation theory. However, due to the nonlinear character of the problems involved and the absence of computers, the discovery of chaos did not take place until 1963. That year Edward Lorenz published his monumental work entitled *Deterministic Nonperiodic Flow*. For the first time it was shown that a system of three nonlinear ordinary differential equations exhibits final states that are nonperiodic. Soon after that the theory of chaos developed to what many consider the third most important discovery in the 20th century after relativity and quantum mechanics.

First, the hidden beauty of chaos was revealed by studying simple nonlinear mathematical models such as the logistic equation, the Hénon map, the Lorenz system, and the Rössler system. Beautiful "strange attractors" that described the final states of these systems were produced and studied, and routes that lead a dynamical system to chaos were discovered.

After that the study of chaos moved to the laboratory. Ingenious experiments were set up, and low-dimensional chaotic behavior was observed. These experiments elevated chaos from being just a mathematical curiosity and established it as a physical reality.

The next step was to search for chaos outside the "controlled" laboratory—in nature. This presented an enormous challenge. Now the scientists had to deal with an "uncontrolled" system whose mathematical formulation was not always known accurately. Up to this point, the existence of low-dimensional chaos in physical systems has not been demonstrated beyond any doubt. Many indications have been presented, but a definite answer has not yet emerged. More work is needed in this area.

The acceptance of a new theory depends on its ability to make predictions. For example, the theory of relativity predicted that light must bend in the presence of a strong gravitational field. This prediction (among others) was soon verified, and the theory became widely accepted. Similar comments can be made about quantum mechanics and other accepted theories. Chaos theory tells us that nonlinear deterministic systems are sensitive to initial conditions and because of that their predictive power is lost very quickly. At the same time we have discovered that processes that appear random may be chaotic, and thus they should be treated as deterministic processes. Would it be possible that the underlying determinism of such processes could be used to improve their otherwise limited predictability? Many argued that if chaos was to make an impact it had to be used to obtain improved predictions. Lately, nonlinear prediction has become a major area of research, and some very exciting results have been reported. Other advances in the theory include the use of chaos to reduce noise in the data.

The book is divided into three parts. The first part (Chapters 1–4) reviews concepts from mathematics, physics, and fractal geometry that we will be using in later chapters. These concepts include stability analysis, conservative systems, and ergodic systems. The second part (Chapters 5–7) presents the fundamentals behind the theory of chaos. Chapter 5 introduces the reader to strange attractors and their characteristics. Chapter 6 provides an overview of bifurcation theory and routes to chaos. Chapter 7 is devoted to the existence of chaos in Hamiltonian, quantum, and partial differential equation (PDE) systems. The third part (Chapters 8–11) is dedicated to the applications of chaos theory. Chapter 8 is concerned with reconstructing the dynamics from observables. Here the "burning" question of the necessary number of points is treated in detail. Chapter 9 is devoted to the evidence of chaos in controlled and uncontrolled systems. Chapter 10 introduces the reader to the rapidly growing area of nonlinear prediction. Chapter 11 gives an introduction to two other important research areas, namely shadowing and noise reduction.

CHAPTER 2

MATHEMATICAL NOTES

1. PROBABILITY DISTRIBUTIONS, EXPECTATIONS, AND MOMENTS

Let us assume that we are throwing darts on a board with the intention to hit the center. As we all know from experience, the darts hit the board (or the wall) at different points. The distance x of each dart from the center can thus assume values in the interval $[0, +\infty]$. The distance x is called a random variable. If we measure the distance for N independent trials, we obtain N values for x. We can find how many of these N values are found in some interval $[0, d)$, how many in the interval $[d, 2d)$, how many in the interval $[2d, 3d)$, and so on. This way we can produce what is called a *frequency distribution* for the variable x. From the frequency distribution we can obtain the *cumulative probability distribution function* $P(X)$, which is defined as

$$P(X) = \mathrm{pr}(x < X)$$

where $\mathrm{pr}(x < X)$ is the probability that the random variable x will assume a value less than some given value X. From the cumulative probability distribution we can then define the *probability density function* $p(X)$ (assuming that $P(X)$ is differentiable):

$$p(X) = \frac{dP(X)}{dX}$$

7

From the above equation we have

$$P(X) = \int_{-\infty}^{X} p(x') \, dx'$$

which defines the probability that x will take on a value between $-\infty$ and X or

$$P(X) = \int_{X=a}^{X=b} p(x') \, dx'$$

the probability that x will take on a value in the interval $[a, b]$, etc.

The *mathematical expectation* or *mean value* of x, denoted $E(x)$ or $\bar{x}$, is defined as

$$\bar{x} = E(x) = \int_{-\infty}^{\infty} Xp(X) \, dX$$

In analogy to the definition from physics, $E(x)$ "represents" the first moment of x (see Fig. 1). The second, third, ..., nth moments of x can be defined accordingly as

$$E(x^2) = \overline{x^2} = \int_{-\infty}^{\infty} X^2 p(X) \, dX$$
$$\vdots$$
$$E(x^n) = \overline{x^n} = \int_{-\infty}^{\infty} X^n p(X) \, dX$$

The *variance* of x is defined as the expected value of the square difference of x about the mean: $\mathrm{Var}(x) = E[(x - \bar{x})^2]$. The *standard deviation* σ is defined as $\sigma = \sqrt{\mathrm{Var}(x)}$.

2. THE AUTOCORRELATION AND SPECTRAL DENSITY FUNCTIONS

A very powerful way to analyze a time series is by calculating the autocorrelation and spectral density function. The autocorrelation function is useful in determining the degree of dependence present in the values of

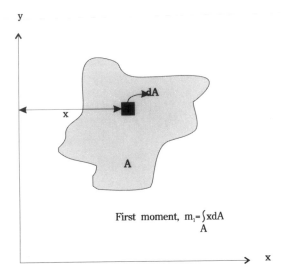

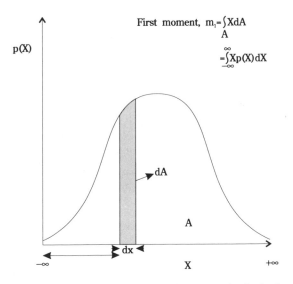

FIGURE 1.   The analogy between the definition of a moment in physics (top) and statistics (bottom).

a time series $x(t)$, separated by an interval called the lag, k. The autocorrelation function $r(k)$ is

$$r(k) = \frac{\text{Cov}(x(t), x(t+k))}{\text{Var}(x(t))}$$

$$= \frac{1}{T-k} \int_0^{T-k} (x(t) - \bar{x})(x(t+k) - \bar{x}) \, dt \bigg/ \text{Var}(x(t))$$

for a continuous process (T is the total length of the time series and $\bar{x}$ is the mean value) or

$$r(k) =$$

$$\frac{\sum_{i=1}^{N-k} x_i x_{i+k} - \frac{1}{N-k} \sum_{i=1}^{N-k} x_{i+k} \sum_{i=1}^{N-k} x_i}{\left[\sum_{i=1}^{N-k} x_i^2 - \frac{1}{N-k}(\sum_{i=1}^{N-k} x_i)^2\right]^{1/2} \left[\sum_{i=1}^{N-k} x_{i+k}^2 - \frac{1}{N-k}(\sum_{i=1}^{N-k} x_{i+k})^2\right]^{1/2}}$$

for a discrete time series of size N.

For a purely random process the autocorrelation function fluctuates randomly about zero, indicating that the process at any certain instance has no "memory" of the past at all (Fig. 2). For a periodic process the autocorrelation function is also periodic, indicating the strong relation between values that repeat over and over again (Fig. 3).

The spectral density function is most useful in isolating periodicities. While the autocorrelation function is useful in analyzing data in the time domain, periodicities often are best delineated by analyzing the data in the frequency domain. According to Fourier analysis, we can represent any irregular periodic time series as a sum of regular sinusoidal waves of various frequencies, amplitudes, and relative phases. For example, in Fig. 4, we represent our "irregular" periodic signal A of period T by the sum of the "regular" signals a, b, and c, which have different amplitudes and phases. In mathematical terms we can thus represent an observable $x(t)$ defined in the interval $0 < t < T$ as

$$x(t) = \sum_{n=0}^{\infty} a_n \cos n\omega t + b_n \sin n\omega t$$

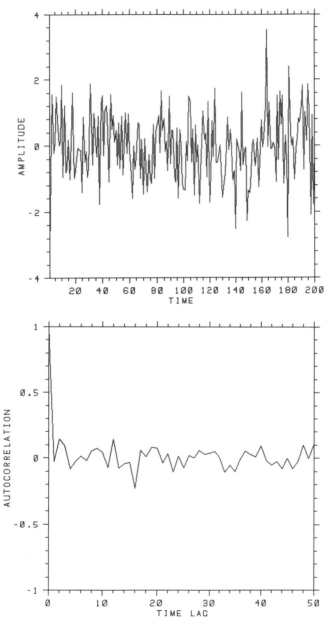

FIGURE 2.   An example of a white noise sequence (top) and its autocorrelation function (bottom).

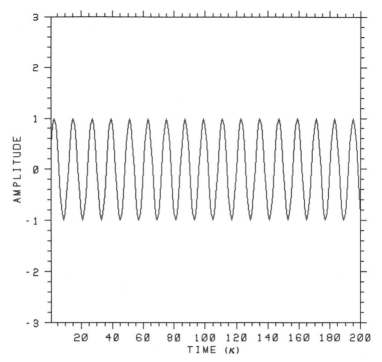

FIGURE 3.   The periodic function $\sin(2\pi k/12)$ as a function of k. The autocorrelation function is identical to the time series.

where $\omega = 2\pi f$ (f is the frequency) and a_n, b_n are a set of coefficients. Note that the energy E contained in $x(t)$ is equal to $\int_0^\infty |x(t)|^2\, dt$.

Once we have done that, we can record the square of the amplitude and the frequency (inverse of periodicity) of each of the component waves in a diagram called the power spectrum. The power spectrum shows the contributions of each of the component waves to the "shaping up" of the irregular signal (Fig. 5). In effect, we now transform the information from the time domain to the frequency domain. Mathematically this means that $x(t)$ is transformed to

$$x(t) = \sum_{n=-\infty}^{\infty} F_n e^{n2\pi ift}$$

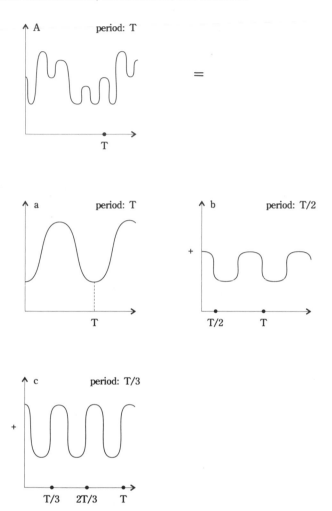

FIGURE 4. The idea behind representing an "irregular" periodic signal A of period T as a sum of a number of regular signals a, b, and c.

or, in integral form,

$$x(t) = \int_{-\infty}^{\infty} F(f)e^{2\pi ift} \, df$$

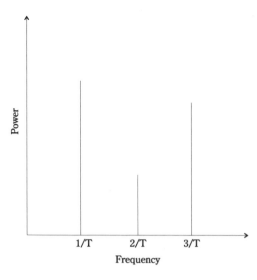

FIGURE 5. If we record the square of the amplitude of signals a, b, and c of Fig. 4 and their frequency (inverse of periodicity), we can obtain a power spectrum that shows the contribution of each of the signals to the total "energy" of signal A.

where $F(f)$ is the Fourier transform of $x(t)$:

$$F(f) = \int_0^T x(t)e^{-2\pi ift}\, dt$$

The total energy is now equal to $\int_{-\infty}^{\infty} |F(f)|^2\, df$ (this is also known as Parseval's theorem). Considering that the contribution of each frequency interval df to the total energy is $|F(f)|^2\, df$, the average power of $x(t)$ in the interval $0 < t < T$ is

$$\frac{1}{T}\int_{-\infty}^{\infty} |F(f)|^2\, df$$

If the irregular periodic signal can be decomposed into n component waves, there are n discrete vertical lines in the power spectrum. In this case the power spectrum is said to be discrete. Obviously, as $n \to \infty$, the spectrum will consist of an infinite number of vertical lines, thus becoming continuous. This is the case when we deal with noise or random signals or, as we will see later, with deterministic but chaotic time series. In effect, the power spectrum partitions the variations in the irregular signal into a number of frequency intervals or bands. For this reason we often refer to the spectral density function, which is simply the amount of variance per interval of frequency. The spectral density $S(f)$ of $x(t)$ is thus defined as

$$S(f) = \frac{1}{T} |F(f)|^2$$

According to a result known as the Weiner-Khintchine theorem, the spectral density function can be expressed as the Fourier transform of the autocorrelation function:

$$S(f) = \int_{-\infty}^{\infty} r(k) e^{-i2\pi fk}\, dk = 2 \int_{0}^{\infty} r(k) \cos(2\pi fk)\, dk$$

For $k = 0$ we have that $r(k) = 1$ and, thus,

$$\int_{-\infty}^{\infty} S(f)\, df = \int_{-\infty}^{\infty} \int_{-\infty}^{\infty} e^{-i2\pi fk}\, dk\, df = 1$$

This is the normalized variance of the series. Since $\int_{-\infty}^{\infty} S(f)\, df = 1$, $S(f)$ has properties of a probability density function and thus can be viewed as a probability density function that gives the contribution to the total normalized variance in the frequency range $\Delta f = f_2 - f_1$ ($\int_{f_1}^{f_2} S(f)\, df$).

In practice, we usually deal with time series that are not continuous but discrete and separated by an amount Δt. In such cases the continuous definition of spectral density will have to be adjusted to take into account the fact that now the variance has to be distributed over a discrete set of frequencies. In such cases the component wave with the highest frequency for which information can be obtained is the wave with frequency $f_N = 1/(2\,\Delta t)$. This is the notorious Nyquist frequency, and it simply tells us that if our data are sampled, say, once every day, then the shortest periodicity that we can possibly have is two days.

For discrete time series the spectral density can be calculated according to

$$S(f) = \Delta t \left[r(0) + 2 \sum_{k=1}^{m-1} r(k) \cos(2\pi k f \Delta t) + r(m) \cos(2\pi m f \Delta t) \right]$$

where m is the maximum number of correlation lags, normally recommended not to exceed 20% of N (the sample size of the time series). The value of $S(f)$ should be evaluated only at frequencies $f = kf_N/m = k/2\Delta tm$. For example, $\Delta t = 1$ yields $f = k/2m$. Thus, since $k_{max} = m$, the highest frequency in this case is 0.5.

For a purely random process the spectral density oscillates randomly about a constant value, indicating that no frequency explains any more of the variance of the sequence than any other frequency (Fig. 6). For a periodic or quasi-periodic, sequence only peaks at certain frequencies exist. At any other frequency the value of the spectral density is zero (Figs. 7 and 8). Basically, both the correlation function and the spectral density function contain the same information (recall that the spectral density is defined as the Fourier transform of the autocorrelation function). The difference is that this information is presented in the time (or spatial) domain by the autocorrelation function and in the frequency domain by

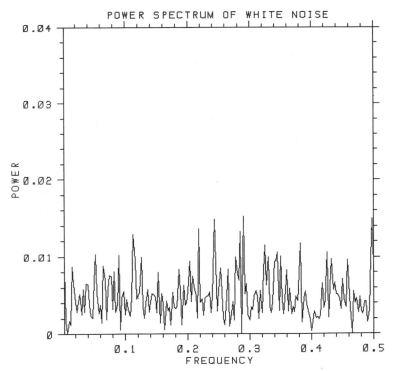

FIGURE 6.   Power spectrum of the white noise sequence in Fig. 2. Note that no particular frequency explains most of the variation of the sequence.

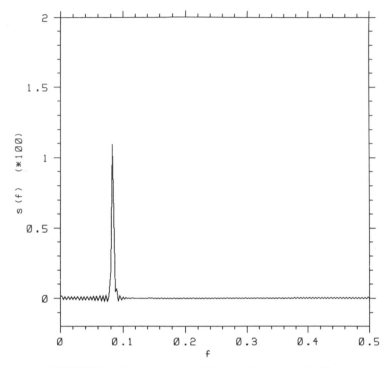

FIGURE 7. Power spectrum of the periodic sequence in Fig. 3.

the spectral density function. Finally, note that the autocorrelation function, by the way it is defined, measures the *linear* dependence between successive values. We will come back to this later.

3. STABILITY ANALYSIS

3.1. Linear Systems

Let us consider a system of two linear, first-order homogeneous differential equations:

$$\dot{x}_1 = a_{11}x_1 + a_{12}x_2$$

$$\dot{x}_1 = a_{21}x_1 + a_{22}x_2 \tag{2.1}$$

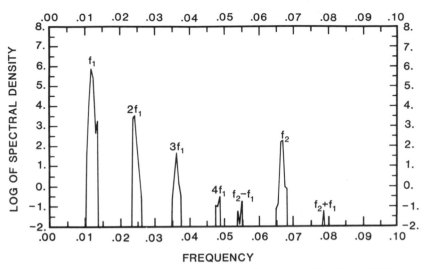

FIGURE 8.   Power spectrum of a quasi-periodic sequence (Tsonis *et al.*[216]).

where the a_{ij} are constants and $\dot{x}$ denotes dx/dt. We can rewrite Eq. (2.1) in vector notation as

$$\dot{\mathbf{x}} = A\mathbf{x} \tag{2.2}$$

where

$$A = \begin{pmatrix} a_{11} & a_{12} \\ a_{21} & a_{22} \end{pmatrix}$$

$$\mathbf{x} = \begin{pmatrix} x_1 \\ x_2 \end{pmatrix} \quad \text{and} \quad \dot{\mathbf{x}} = \begin{pmatrix} \dot{x}_1 \\ \dot{x}_2 \end{pmatrix}$$

The equilibrium of system (2.1) can be found if we set $\dot{\mathbf{x}} = 0$ or $A\mathbf{x} = 0$. Therefore, if A is nonsingular ($A \neq 0$), the only equilibrium state is $\mathbf{x} = 0$ ($x_1 = 0$ and $x_2 = 0$ in this example).

As in the case of a scalar (nonvector) first-order equation $\dot{x} = ax$, we also assume that a solution of Eq. (2.2) is of the form

$$\mathbf{x}(t) = \mathbf{c} e^{\lambda t} \tag{2.3}$$

where λ is a scalar and $\mathbf{c}$ is a nonzero vector $\begin{pmatrix} c_1 \\ c_2 \end{pmatrix}$.

Equation (2.3) can be rewritten as

$$\begin{pmatrix} x_1(t) \\ x_2(t) \end{pmatrix} = e^{\lambda t} \begin{pmatrix} c_1 \\ c_2 \end{pmatrix} \tag{2.4}$$

which when substituted into (2.1) gives

$$A\mathbf{c} = \lambda\mathbf{c} \tag{2.5}$$

A nontrivial solution to Eq. (2.5) for a given λ is called an eigenvector, and λ is called the eigenvalue. Because we require a nontrivial (nonzero) solution, it is necessary that

$$\mathrm{Det}(A - \lambda I) = 0 \tag{2.6}$$

Equation (2.6) is called the determinant equation, and it can be written as

$$\mathrm{Det}\begin{pmatrix} a_{11} - \lambda & a_{12} \\ a_{21} & a_{22} - \lambda \end{pmatrix} = 0$$

or

$$(a_{11} - \lambda)(a_{22} - \lambda) - a_{12}a_{21} = 0$$

or

$$\lambda^2 + a_{11}a_{22} - a_{12}a_{21} - \lambda(a_{11} + a_{22}) = 0$$

or

$$\lambda^2 - \lambda\, \mathrm{Trace}\, A + \mathrm{Det}A = 0 \tag{2.7}$$

The quadratic equation (2.7) is called the characteristic equation. Its solutions λ_1 and λ_2 are eigenvalues which are either both real or both complex (complex conjugate). Once the eigenvalues have been determined, solution of Eq. (2.5) gives the corresponding eigenvectors

$$\mathbf{c}_1 = \begin{pmatrix} c_{11} \\ c_{12} \end{pmatrix} \quad \text{and} \quad \mathbf{c}_2 = \begin{pmatrix} c_{21} \\ c_{22} \end{pmatrix}$$

Assuming that $\lambda_1 \neq \lambda_2$, it follows that $\mathbf{c}_1$ and $\mathbf{c}_2$ are two linearly independent vectors in R^2. Thus, both $\mathbf{c}_1 e^{\lambda t}$ and $\mathbf{c}_2 e^{\lambda t}$ are solutions to Eq. (2.3), and since the original system (Eq. (2.1)) is linear, any linear combination of these two solutions will also constitute a solution. Thus, we may write the general solution of (2.1) as

$$\mathbf{x}(t) = a_1 e^{\lambda_1 t} \mathbf{c}_1 + a_2 e^{\lambda_2 t} \mathbf{c}_2 \tag{2.8}$$

Suppose now that λ_1, λ_2 are both real. If they are both negative, then $\mathbf{x}(t) \to 0$ as $t \to \infty$ independently of the initial condition $\mathbf{x}(0)$. Thus, in this case the system is attracted to the equilibrium state no matter where it starts from. We say that the equilibrium state is *asymptotically stable.* The stable equilibrium state is called a fixed point or a node or an elliptic point. If λ_1 and λ_2 are real positive, then we conclude that, as $t \to \infty$, $\mathbf{x}(t) \to \infty$. In this case, no matter what the initial condition, the system *will not* approach the equilibrium state. As a matter of fact, even if it starts very close to the equilibrium state it will have to go to infinity. In this case we say that the origin repels all initial states. It is, therefore, unstable. If $\lambda_1 < 0 < \lambda_2$, then we find that the contribution of λ_1 tries to make the final-state approach the equilibrium state, whereas the contribution of λ_2 tries to repel final states from the equilibrium state. The linear combination of these two motions leads to evolutions that appear to approach the equilibrium state and then move away. In this case we call the unstable equilibrium state a *saddle* (or a hyperbolic point).

If λ_1 and λ_2 are complex with $\lambda_i = \alpha + \beta i$, then $\mathbf{x}(t) = e^{\alpha t}(\mathbf{k}_1 \cos \beta t + \mathbf{k}_2 \sin \beta t)$, where $\mathbf{k}_1$ and $\mathbf{k}_2$ are appropriate vectors. If α is negative, then $\mathbf{x}(t) \to 0$ as $t \to \infty$, as again the equilibrium state is asymptotically stable. If α is positive, then $\mathbf{x}(t) \to 0$ as $t \to \infty$, and the equilibrium state is unstable. If $\alpha = 0$, then the solution is periodic with the periodicity being determined by the initial condition (for each different initial condition there exists a distinct periodic solution). In this case we say that we have neutral stability. The equilibrium point is called a center or a vortex (often this state is also referred to as an elliptic point).

In summary, a homogeneous system of n first-order ordinary differential equations is asymptotically stable if the real parts of its eigenvalues are negative and it is unstable otherwise. Here we may recall Eq. (2.7) and that

$$\lambda_{1,2} = \frac{\text{Trace } A - \sqrt{(\text{Trace } A)^2 - 4 \text{ Det } A}}{2}$$

Now we can easily prove that

$$\lambda_1 \lambda_2 = \text{Det}A$$

$$\lambda_1 + \lambda_2 = \text{Trace } A \qquad (2.9)$$

If we combine Eq. (2.9) and our conclusions from above, we can further conclude that for $n = 2$ the equilibrium state of the system is asymptotically stable if and only if $\text{Det}A > 0$ and $\text{Trace } A < 0$. In any other case the equilibrium state is unstable. Figure 9 summarizes all possible equilibrium states and their stability.

Example. Consider the system

$$\dot{x}_1 = x_1 + 2x_2$$
$$\dot{x}_2 = 3x_1 + 2x_2$$

We can rewrite this system as $\dot{\mathbf{x}} = A\mathbf{x}$, where

$$A = \begin{pmatrix} 1 & 2 \\ 3 & 2 \end{pmatrix}$$

The eigenvalues of A can be found via the equation

$$|A - \lambda I| = 0$$

where

$$I = \begin{pmatrix} 1 & 0 \\ 0 & 1 \end{pmatrix}$$

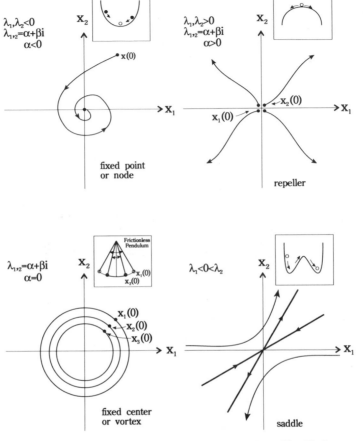

FIGURE 9.   Summary of all possible equilibrium states and their stability. The insets represent simple physical systems appropriate to each case.

Since

$$|A - \lambda I| = \begin{vmatrix} 1 - \lambda & 2 \\ 3 & 2 - \lambda \end{vmatrix} = \lambda^2 - 3\lambda - 4 = (\lambda - 4)(\lambda + 1)$$

the eigenvalues of A are $\lambda_1 = -1$ and $\lambda_2 = 4$.

To find the corresponding linearly independent eigenvectors

$$\mathbf{c}_1 = \begin{pmatrix} c_{11} \\ c_{21} \end{pmatrix} \quad \text{and} \quad \mathbf{c}_2 = \begin{pmatrix} c_{12} \\ c_{22} \end{pmatrix}$$

we seek the solution of

$$(A - \lambda_1 I) = 0 \quad \text{and} \quad (A - \lambda_2 I) = 0$$

From the first of the above equations we have

$$(A - \lambda_1 I)\mathbf{c}_1 = \begin{pmatrix} 2 & 2 \\ 3 & 3 \end{pmatrix} \begin{pmatrix} c_{12} \\ c_{21} \end{pmatrix} = \begin{pmatrix} 2c_{12} + 2c_{21} \\ 3c_{12} + 3c_{21} \end{pmatrix}$$

For this to be zero we must have $c_{12} = -c_{21}$. We may choose $c_{12} = 1$ and $c_{21} = -1$. Thus,

$$\mathbf{c}_1 = \begin{pmatrix} 1 \\ -1 \end{pmatrix}$$

For $\lambda_2 = 4$ we have

$$(A - \lambda_2 I)\mathbf{c}_2 = \begin{pmatrix} -3 & 2 \\ 3 & -2 \end{pmatrix} \begin{pmatrix} c_{12} \\ c_{22} \end{pmatrix} = \begin{pmatrix} -3c_{12} + 2c_{22} \\ 3c_{12} - 2c_{22} \end{pmatrix}$$

For this to be zero we must have $c_{12} = (\frac{2}{3})c_{22}$. We may choose $c_{22} = 3$, which yields $c_{12} = 2$ and, thus,

$$\mathbf{c}_2 = \begin{pmatrix} 2 \\ 3 \end{pmatrix}$$

Let us now define the matrix

$$T = \begin{pmatrix} 1 & 2 \\ -1 & 3 \end{pmatrix}$$

whose columns are the eigenvalues c_1 and c_2. This matrix has inverse

$$T^{-1} = \begin{pmatrix} 3/5 & -2/5 \\ 1/5 & 1/5 \end{pmatrix}$$

One can verify that $TT^{-1} = I$ and

$$T^{-1}AT = \begin{pmatrix} -1 & 0 \\ 0 & 4 \end{pmatrix}$$

Therefore we find that $T^{-1}AT$ is a diagonal matrix whose entries are the eigenvalues of matrix A. *Remember this result for later!* We can also see that $\text{Det}A = -4 = \lambda_1\lambda_2$ and that Trace $A = 3 = \lambda_1 + \lambda_2$.

3.2. Nonlinear Systems

When we are dealing with a nonlinear system of n first-order ordinary differential equations (ODEs), an analytic solution is usually not obtainable, and the process described above cannot be applied to study its stability. In this case the system is expressed as

$$\begin{aligned}
\dot{x}_1 &= f_1(x_1, x_2, \ldots, x_n) \\
\dot{x}_2 &= f_2(x_1, x_2, \ldots, x_n) \\
&\vdots \\
\dot{x}_n &= f_n(x_1, x_2, \ldots, x_n)
\end{aligned}$$

(2.10)

where f_1, f_2, and f_n are nonlinear functions of all or some of the variables $x_1, x_2, \ldots, x_n$. The equilibrium state(s) of the system is (are) again defined by setting $\dot{x}_i = 0$, $i = 1, 2, \ldots, n$.

To address the issue of stability in this case, we have to proceed with an analysis which investigates the properties of the system for $x_1 = \bar{x}_1 + x_1'$, $x_2 = \bar{x}_2 + x_2', \ldots, x_n = \bar{x}_n + x_n'$, where $x_i' \ll 0$ indicate small deviations from an equilibrium state $(\bar{x}_1, \bar{x}_2, \ldots, \bar{x}_n)$. We can rewrite the system of Eqs. (2.10) as

$$\frac{d(\bar{x}_1 + x'_1)}{dt} = f_1(\bar{x}_1 + x'_1, \bar{x}_2 + x'_2, \ldots, \bar{x}_n + x'_n)$$

$$\frac{d(\bar{x}_2 + x'_2)}{dt} = f_2(\bar{x}_1 + x'_1, \bar{x}_2 + x'_2, \ldots, \bar{x}_n + x'_n) \qquad (2.11)$$

$$\vdots$$

$$\frac{d(\bar{x}_n + x'_n)}{dt} = f_n(\bar{x}_1 + x'_1, \bar{x}_2 + x'_2, \ldots, \bar{x}_n + x'_n)$$

We now proceed by simplifying the above system of differential equations by setting all *nonlinear* terms involving fluctuations on the right side equal to zero. This way we effectively replace $f_1(\bar{x}_1 + x'_1, \bar{x}^2 + x'_2, \ldots, \bar{x}_n + x'_n)$ by $f'_1(x'_1, x'_2, \ldots, x'_n) = a_{11}x'_1 + a_{12}x'_2 + \cdots + a_{1n}x'_n$, etc. Considering that $d\bar{x}/dt = 0$, we can rewrite system (2.11) as

$$\dot{x}'_1 = a_{11}x'_1 + a_{12}x'_2 + \cdots + a_{1n}x'_n$$

$$\dot{x}'_2 = a_{21}x'_1 + a_{22}x'_2 + \cdots + a_{2n}x'_n$$

$$\vdots \qquad\qquad (2.12)$$

$$\dot{x}'_n = a_{n1}x'_1 + a_{n2}x'_2 + \cdots + a_{nn}x'_n$$

This is a linear system of first-order differential equations describing the evolution of the *fluctuations* $x'_1, x'_2, \ldots, x'_n$ about an equilibrium state $\bar{x}_1, \bar{x}_2, \ldots, \bar{x}_n$. Obviously, if the fluctuations grow in time, then the system is driven away from the equilibrium state and it is unstable. Otherwise it is stable. We stress that we *do not* really investigate the *global* properties of the system by its *local* properties (those very near the equilibrium states).

We may represent system (2.12) in vector form

$$\dot{\mathbf{x}}' = A\mathbf{x}' \qquad (2.13)$$

where

$$A = \begin{pmatrix} a_{11} & a_{12} & \cdots & a_{1n} \\ a_{21} & a_{22} & \cdots & a_{2n} \\ \vdots & \vdots & & \vdots \\ a_{n1} & a_{n2} & \cdots & a_{nn} \end{pmatrix}$$

The stability of Eq. (2.13) is, as discussed before, determined by the eigenvalues λ_i, defined by the equation

$$\text{Det}(A - \lambda I) = 0$$

We may recognize here that the matrix A is the Jacobian matrix of $\mathbf{f}$ evaluated at $\bar{\mathbf{x}}$. This is a direct result of the fact that, according to Taylor's theorem, a nonlinear function $f(x_1, x_2, \ldots, x_n)$ is equal to

$$f(x_1, x_2, \ldots, x_n)$$

$$= f(\bar{x}_1, \bar{x}_2, \ldots, \bar{x}_n) + x_1' \frac{\partial f}{\partial x_1}\bigg|_{x_1=\bar{x}_1, x_2=\bar{x}_2, \ldots, x_n=\bar{x}_n}$$

$$+ x_2' \frac{\partial f}{\partial x_2}\bigg|_{x_1=\bar{x}_1, x_2=\bar{x}_2, \ldots, x_n=\bar{x}_n} + \cdots + x_n' \frac{\partial f}{\partial x_n}\bigg|_{x_1=\bar{x}_1, x_2=\bar{x}_2, \ldots, x_n=\bar{x}_n}$$

$$+ \text{ higher-order terms}$$

At equilibrium $f(\bar{x}_1, \bar{x}_2, \ldots, \bar{x}_n) = 0$, and by neglecting all higher-order terms we obtain

$$f(x_1, x_2, \ldots, x_n) = x_1' \frac{\partial f}{\partial x_1}\bigg|_{x_1=\bar{x}_1, x_2=\bar{x}_2, \ldots, x_n=\bar{x}_n}$$

$$+ \cdots + x_n' \frac{\partial f}{\partial x_n}\bigg|_{x_1=\bar{x}_1, x_2=\bar{x}_2, \ldots, x_n=\bar{x}_n}$$

Since $\dot{x} = f(x_1, x_2, \ldots, x_n)$, for a nonlinear system of n first-order ODEs we have

$$\dot{\mathbf{x}} = A'\mathbf{x}'$$

or (since $\dot{x} = \dot{\bar{x}} + \dot{x}'$ and $\dot{\bar{x}} = 0$)

$$\dot{\mathbf{x}}' = A'\mathbf{x}' \tag{2.14}$$

where

$$
A' = \begin{pmatrix}
\dfrac{\partial f_1}{\partial x_1} & \dfrac{\partial f_1}{\partial x_2} & \cdots & \dfrac{\partial f_1}{\partial x_n} \\[2ex]
\dfrac{\partial f_2}{\partial x_1} & \dfrac{\partial f_2}{\partial x_2} & \cdots & \dfrac{\partial f_2}{\partial x_n} \\[2ex]
\vdots & \vdots & & \vdots \\[2ex]
\dfrac{\partial f_n}{\partial x_1} & \dfrac{\partial f_n}{\partial x_2} & \cdots & \dfrac{\partial f_n}{\partial x_n}
\end{pmatrix}
\tag{2.15}
$$

Since Eq. (2.14) is identical to Eq. (2.13), it follows that $A = A'$. This result provides us in many cases with an easier way to investigate stability of equilibrium states of a system of differential equations. Note that it is always the eigenvalues computed from the equation $\mathrm{Det}(A - \lambda I) = 0$ [or $\mathrm{Det}(A' - \lambda I) = 0$] that determine the stability of a system. Only if A (or A') is a 2×2 matrix may we use the shortcut of simply finding $\mathrm{Det}A$ and Trace A and requiring for stability that $\mathrm{Det}A > 0$ and Trace $A < 0$.

Example. Let us now put the theory into practice by considering an example. Our example is a set of ODEs describing the motion of a pendulum:

$$
\dot{x}_1 = x_2
$$

$$
\dot{x}_2 = -\frac{g}{l}\sin x_1 - rx_2
\tag{2.16}
$$

In accordance with previous notation, $f_1(x_1, x_2) = x_2$ and $f_2(x_1, x_2) = -(g/l)\sin x_1 - rx_2$. The variables x_1 and x_2 are the angle from the vertical and the velocity. The constants g, l, and r are the acceleration of gravity, the length of the pendulum, and a constant related to the mass of the pendulum, respectively.

The equilibrium points can be found from the equations

$$
x_2 = 0 \qquad -\frac{g}{l}\sin x_1 - rx_2 = 0
$$

This system has two solutions. The first is $\bar{x}_2 = 0$, $\bar{x}_1 = 0$, and the second is $\bar{x}_2 = 0$, $\bar{x}_1 = \pi$. We thus have two possible equilibrium states (Fig. 10).

Let us now examine the stability of these states in the presence of small fluctuations in position x_1' and velocity x_2'. These fluctuations will take the system to some state $x_1 = \bar{x}_1 + x_1'$ and $x_2 = \bar{x}_2 + x_2'$. If we substitute these states into system (2.16), we have

$$\frac{d(\bar{x}_1 + x_1')}{dt} = \bar{x}_2 + x_2'$$

$$\frac{d(\bar{x}_2 + x_2')}{dt} = -\frac{g}{l}\sin(\bar{x}_1 + x_1') - r(\bar{x}_2 + x_2')$$

Taking first $\bar{x}_1 = 0$, $\bar{x}_2 = 0$, we arrive at

$$\dot{x}_1' = x_2'$$

$$\dot{x}_2' = -\frac{g}{l}\sin x_1' - rx_2' \qquad (2.17)$$

Since $x_1' \ll 0$, we may assume that $\sin x_1' = x_1'$, and thus Eq. (2.17) becomes

$$\dot{x}_1' = x_2'$$

$$\dot{x}_2' = -\frac{g}{l}x_1' - rx_2' \qquad (2.18)$$

System (2.18) is now linear. The process we have followed has reduced the original nonlinear system to a linear system for the deviations x_1' and x_2'.

We can rewrite (2.18) in vector notation as

$$\dot{\mathbf{x}} = A\mathbf{x}$$

where

$$\dot{\mathbf{x}} = \begin{pmatrix} \dot{x}_1' \\ \dot{x}_2' \end{pmatrix} \quad \mathbf{x} = \begin{pmatrix} x_1' \\ x_2' \end{pmatrix} \quad \text{and} \quad A = \begin{pmatrix} 0 & 1 \\ -g/l & -r \end{pmatrix}$$

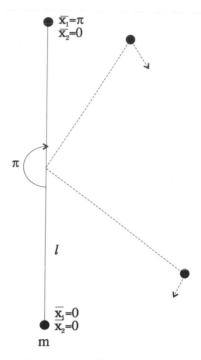

FIGURE 10. Equilibrium states of a pendulum of mass m and length l.

Thus, the equilibrium state $\bar{x}_1 = 0$, $\bar{x}_2 = 0$ is asymptotically stable since $\text{Det}A = g/l > 0$ and Trace $A = -r < 0$. If we were examining the stability of the system near the equilibrium state $\bar{x}_1 = \pi$, $\bar{x}_2 = 0$, then

$$A = \begin{pmatrix} 0 & 1 \\ \dfrac{g}{l} & -r \end{pmatrix}$$

which implies that $\text{Det}A = -g/l < 0$ and Trace $A = -r < 0$. Thus, this equilibrium position of the pendulum, as anticipated, is unstable.

3.3. Stability of maps

The systems we have considered up to this point are described by a set of rules (differential equations). Such systems are often called *dynamical systems*. Since the solution of a set of differential equations is continuous, these systems are called *flows*. A dynamical system, however, does not have to be described by a set of differential equations. Many dynamical systems are described by a set of difference equations, and they are often referred to as *maps*. For example, the dynamical system

$$x_{n+1} = 1 - ax_n^2 + y_n$$

$$y_{n+1} = bx_n \qquad (2.19)$$

is a two-dimensional map called the Hénon map,[101] while the dynamical system

$$x_{n+1} = \mu x_n(1 - x_n) \qquad (2.20)$$

is a one-dimensional system called the logistic equation.[146]

Stability analysis on maps is very similar to stability analysis on flows. Let us proceed with an example and consider the stability of the Hénon map. The equilibrium points (or fixed points) are found by assuming that $x_{n+1} = x_n = \bar{x}$ and $y_{n+1} = y_n = \bar{y}$ (here instead of $\dot{x} = 0$ we assume that at equilibrium $x_{n+1} - x_n = 0$). From Eq. (2.19) we have that, at equilibrium,

$$\bar{x} = 1 - a\bar{x}^2 + \bar{y}$$

$$\bar{y} = b\bar{x} \qquad (2.21)$$

or

$$\bar{x} = 1 - a\bar{x}^2 + b\bar{x}$$

or

$$a\bar{x}^2 + (1 - b)\bar{x} - 1 = 0$$

This is a second-degree algebraic equation, and the two solutions are

$$\bar{x}_{1,2} = \frac{(b - 1) \pm \sqrt{(b - 1)^2 + 4a}}{2a}$$

Thus, the two fixed points are

$$\left(\bar{x}_1 = \frac{(b-1) + \sqrt{(b-1)^2 + 4a}}{2a}, \bar{y}_1 = b\bar{x}_1 \right)$$

$$\left(\bar{x}_2 = \frac{(b-1) - \sqrt{(b-1)^2 + 4a}}{2a}, \bar{y}_2 = b\bar{x}_2 \right)$$

We are now interested in examining the stability of the map about the equilibrium states. As with flows, we proceed by investigating the properties of the map in the presence of small fluctuations about an equilibrium state, that is, for $x = \bar{x} + x', y = \bar{y} + y'$. In this case we may rewrite (2.19) as

$$\bar{x} + x'_{n+1} = 1 - a(\bar{x} + x'_n)^2 + \bar{y} + y'_n$$

$$\bar{y} + y'_{n+1} = b\bar{x} + b\bar{x}'_n$$

or as

$$\bar{x} + x'_{n+1} = 1 - a\bar{x}^2 - 2a\bar{x}x'_n - ax'^2_n + \bar{y} + y'_n$$

$$\bar{y} + y'_{n+1} = b\bar{x} + bx'_n$$

Neglecting all terms involving fluctuations higher than first order and taking into account Eq. (2.21), we arrive at

$$x'_{n+1} = -2a\bar{x}x'_n + y'_n$$

$$y'_{n+1} = bx'_n$$

This is now a set of two linear first-order difference equations. It can be written in vector form as

$$\begin{pmatrix} x'_{n+1} \\ y'_{n+1} \end{pmatrix} = A \begin{pmatrix} x'_n \\ y'_n \end{pmatrix} \tag{2.22}$$

where

$$A = \begin{pmatrix} -2a\bar{x} & 1 \\ b & 0 \end{pmatrix}$$

Having expressed our original map in the form of Eq. (2.22), we can then employ our previously derived conditions for stability. Thus, the Hénon map is stable if Det$A = -b > 0$ and Trace $A = -2a\bar{x} < 0$, and unstable otherwise. Here, again, A can be recognized as the Jacobian of $\mathbf{f}$ at $\bar{\mathbf{x}}$ for $f_1(x, y) = 1 - ax^2 + y$ and $f_2(x, y) = bx$.

CHAPTER 3

PHYSICS NOTES

1. CONSERVATIVE VERSUS DISSIPATIVE DYNAMICAL SYSTEMS

Any system whose evolution from some initial state is dictated by a set of rules is called a *dynamical system*. When these rules are a set of differential equations, the system is called a flow, because their solution is continuous in time. When the rules are a set of discrete difference equations, the system is referred to as a map. The evolution of a dynamical system is best described in its phase space, a coordinate system whose coordinates are all the variables that enter the mathematical formulation of the system (i.e., the variables necessary to completely describe the state of the system at any moment). To each possible state of the system there corresponds a point in phase space. If the system in question is just a point particle of mass m, then its state at any given moment is completely described by its speed v and position r (relative to some fixed point). Thus, its phase space is two dimensional with coordinates v and r or $p = mv$ and $q = r$, as in the common Newtonian notation. If instead we were dealing with a cloud of N particles, each of mass m, the phase space would be $2N$-dimensional with coordinates $p_1, p_2, \ldots, p_n, q_1, q_2, \ldots, q_n$. Note that N indicates the number of independent positions or momenta or the number of degrees of freedom. In classical mechanics the total energy E of a dynamical system is related to the equations of motion via the *Hamiltonian H*:

$$H = H(q, p) = E_{\text{kinetic}} + E_{\text{potential}}$$

The central point here is that if $E(p, q)$ remains constant in time, then once the Hamiltonian is known the motion of the system can be completely determined.

Let us again assume that our system is just one particle of mass m. Then

$$H = \frac{1}{2} mv^2 + V(q) = \frac{p^2}{2m} + V(q) \tag{3.1}$$

where $V(q)$ is a suitable expression for the potential energy. From Eq. (3.1) it follows that

$$\frac{\partial H}{\partial p} = \frac{p}{m} = v = \frac{dr}{dt} = \frac{dq}{dt} \tag{3.2}$$

and that

$$\frac{\partial H}{\partial q} = \frac{\partial V(q)}{\partial q} = -F = -ma = -\frac{dp}{dt} \tag{3.3}$$

We can generalize Eqs. (3.2) and (3.3) by considering a system of N particles:

$$\frac{dq_i}{dt} = \frac{\partial H}{\partial p_i}$$
$$\frac{dp_i}{dt} = -\frac{\partial H}{\partial q_i} \tag{3.4}$$

From Eqs. (3.2) and (3.3) or (3.4) it follows that if $H = f(q, p)$ [i.e., if $E(p, q)$ is not a function of time] then the motion of the system can be completely determined. By definition, dynamical systems whose Hamiltonian does not vary with time are called *conservative systems*. Otherwise they are called *dissipative systems*.

Conservative systems have interesting properties, the most important of which is the conservation of volumes in phase space. The conservation of mass of a fluid in motion is expressed analytically by the continuity equation

$$\frac{\partial \rho}{\partial t} + \text{div}\rho\mathbf{v} = 0 \tag{3.5}$$

where ρ signifies the density of the fluid at time t and $\mathbf{v}$ is the velocity of the fluid at the space point under consideration. Equation (3.5) and its foregoing interpretations hold for the motion of a "fluid" made of N phase-space points, provided that $\rho = \rho(q, p, t)$ now stands for the density-in-phase and $\mathbf{v}$ for the "velocity" of phase points [i.e., the $2N$-dimensional vector with components $q(t)$ and $p(t)$]. The term $\text{div}\mathbf{v}$ is defined as

$$
\begin{aligned}
\text{div}\rho\mathbf{v} &= \sum_{i=1}^{2N} \left(\frac{\partial \rho \dot{q}_i}{\partial q_i} + \frac{\partial \rho \dot{p}_i}{\partial p_i} \right) \\
&= \sum_{i=1}^{2N} \left(\dot{q}_i \frac{\partial \rho}{\partial q_i} + \dot{p}_i \frac{\partial \rho}{\partial p_i} \right) + \rho \sum_{i=1}^{2N} \left(\frac{\partial \dot{q}_i}{\partial q_i} + \frac{\partial \dot{p}_i}{\partial p_i} \right)
\end{aligned}
\tag{3.6}
$$

The second term on the right side of Eq. (3.6) is equal to $\rho\, \text{div}\mathbf{v}$. If we recall now that for conservative systems $\partial H/\partial p_i = \dot{q}_i$ and $\partial H/\partial q_i = -\dot{p}_i$, we find that $\text{div}\mathbf{v} = 0$. Thus, Eq. (3.5) reduces to

$$\frac{\partial \rho}{\partial t} + \sum_{i=1}^{2N} \left(\dot{q}_i \frac{\partial \rho}{\partial q} + \dot{p}_i \frac{\partial \rho}{\partial p_i} \right) = 0 \tag{3.7}$$

Equation (3.7), known as Liouville's theorem, simply says that, in a $2N$-dimensional coordinate system $p_1, p_2, \ldots, p_n, q_1, q_2, \ldots, q_n$,

$$\frac{d\rho}{dt} = 0$$

Thus, since $m = \rho V$ (where V denotes volume), we find that for conservative systems $dV/dt = 0$; i.e., volumes in phase space are conserved.

Example. Consider the dynamical system

$$
\begin{aligned}
\dot{x} &= x \\
\dot{y} &= -y
\end{aligned}
\tag{3.8}
$$

The phase space is two dimensional, and its coordinates are x and y. The Hamiltonian of the system can be found from

$$\frac{\partial H}{\partial x} = -\frac{dy}{dt} = y$$

which implies that

$$H(x, y) = yx + c_1 \tag{3.9}$$

and from

$$\frac{\partial H}{\partial y} = \frac{dx}{dt} = x$$

which implies that

$$H(x, y) = xy + c_2 \tag{3.10}$$

From Eqs. (3.9) and (3.10) we conclude that $H = xy$. The solutions of system (3.8) are

$$x = x_0 e^t$$
$$y = y_0 e^{-t} \tag{3.11}$$

where x_0 and y_0 define some initial state of the system at $t = 0$. Thus, $H = xy = x_0 y_0$ is not a function of time.

Let us now consider the evolution of four initial conditions that form a square in phase space, say $x_1 y_1$, $x_2 y_1$, $x_1 y_2$, and $x_2 y_2$ (see Fig. 11). Let us assume that after some time these initial conditions have evolved to $x_1' y_1'$, $x_2' y_1'$, $x_1' y_2'$, $x_2' y_2'$. The area enclosed by the initial conditions is $A = (x_2 - x_1)(y_2 - y_1)$, and the area some time later is $A' = (x_2' - x_1')(y_2' - y_1')$. According to Eqs. (3.11), $A = A'$. Thus, the initial volume in phase space is conserved (although it may be deformed).

The trajectory in phase space is defined according to the equation

$$xy = x_0 y_0 = H(x, y) = E = \text{constant}$$

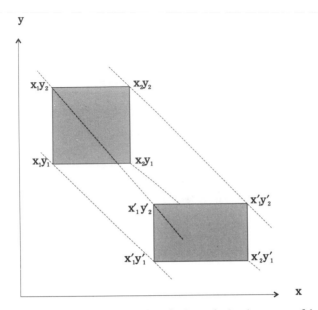

FIGURE 11.   Illustration of the conservation of volumes in the phase space of the dynamical system expressed by Eq. (3.8). The volumes may be deformed, but they remain the same.

Thus, for conservative systems the trajectories are constant-energy trajectories. Since each initial condition in phase space defines a unique constant-energy trajectory, the energy surface is the complete phase space.

If instead of the system (3.8) we had considered the system

$$\dot{x} = -x$$
$$\dot{y} = -y$$

$$(3.12)$$

then it easily follows that $H = -x_0 y_0 e^{-2t}$ and that $A \neq A'$ (see Fig. 12), which means that the initial volume decreases with time.

There are several interesting implications of the definition of conservative and dissipative systems. The frictionless pendulum, for example, is a conservative system. We start it from some initial condition θ_1, and it swings back and forth indefinitely with a maximum angle corresponding to θ_1. If instead we choose an initial condition θ_2, the pendulum will swing back and forth indefinitely with the maximum angle now being θ_2, and so on.

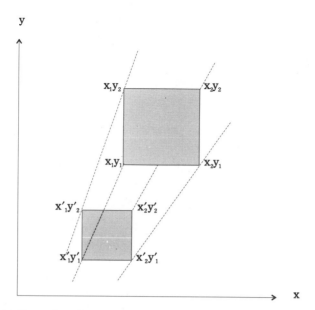

FIGURE 12.   Illustration of the dissipation of volumes in the phase space of the dynamical system expressed by Eq. (3.12). The volumes decrease as time increases.

What we learn from this example is that such a system never forgets a change. The pendulum was swinging between θ_1 and $-\theta_1$; we then introduced a change (or a perturbation), and the pendulum begun an endless swing between θ_2 and $-\theta_2$. If we introduce friction, the system becomes dissipative. We may start the pendulum from an initial condition θ_0, but it will eventually come to a stop at the resting position. In fact, it will come to a stop at the resting position no matter what θ_0 is. In this case we say that the system does not remember the disturbance forever. This is the property of asymptotic stability and is directly related to *irreversibility,* a property of most processes in nature.

Irreversible processes are one-way evolutions. In mathematical terms a process is reversible if it is indistinguishable when time is reversed. Take, for example, Newton's law $m(d^2r/dt^2) = F$. This law describes the motion of an object in the absence of friction. As time increases, an observer will detect a certain orbit. If at some instance we reverse time, so that now $t' = -t$, we find that $m(d^2r/dt^2) = m[d^2r/d(-t)^2] = F$, which means that the observer detects the same property (orbit in this case) but the direction of flow is reversed.

Let us now consider Fourier's equation $\partial T/\partial t = k\nabla^2 T$ ($k > 0$), which describes the diffusion of heat. According to this equation, if we start with a uniform temperature distribution and introduce a fluctuation, this fluctuation "spreads" out and is eventually absorbed. If we reverse time, then this equation is written $\partial T/\partial t = -k\nabla^2 T$. Accordingly, the fluctuation is not absorbed but amplifies continuously. This is obviously not equivalent to the example described before by Newton's equation, and is definitely not what happens in nature. Therefore, we should not expect to easily find reversible processes or conservative systems in nature. We should expect that most often we will have to deal with dissipative systems.

If we now make the connection with some results from the previous chapter, we can derive the necessary conditions for a map or a flow to be a conservative or a dissipative system. For maps, by regarding a set of perturbations as defining some initial volume in phase space, we can see from Eq. (2.22) that this volume will not grow or decay (i.e., it will be conserved) if $|\mathrm{Det}A| = 1$. Thus, a certain map corresponds to a conservative system if $|\mathrm{Det}A| = 1$ and to a dissipative system if $|\mathrm{Det}A| < 1$. Note, however, that for a flow, whether volumes expand or contract or remain the same is given by the trace of A and not by the determinant. Next we see why.

For a flow the evolution of a perturbation is given by Eq. (2.13). This equation has solution

$$\mathbf{x}'(t) = e^{tA}\mathbf{x}'(0) \tag{3.13}$$

where $\mathbf{x}'(0)$ is the initial vector. Assuming that A has distinct eigenvalues $\lambda_1, \lambda_2, \ldots, \lambda_n$, one can find a matrix U such that

$$U^{-1}AU = D$$

where D is diagonal:

$$D = \begin{bmatrix} \lambda_1 & & & \\ & \lambda_2 & & \\ & & \ddots & \\ & & & \lambda_n \end{bmatrix}$$

Then one can write A in the form

$$A = UDU^{-1}$$

Using the multiplication theorem, we have that

$$\mathrm{Det}A = (\mathrm{Det}U)(\mathrm{Det}D)(\mathrm{Det}U^{-1}) = \mathrm{Det}D$$
$$= \lambda_1\lambda_2 \cdots \lambda_n$$

Similarly,

$$\mathrm{Trace}\ A = \mathrm{Trace}\ D = \lambda_1 + \lambda_2 + \cdots \lambda_n$$

In fact, we can generalize the above to consider not just A but any function $f(A)$. By doing this, we find

$$\mathrm{Det}[f(A)] = \mathrm{Det}[f(D)] = f(\lambda_1)f(\lambda_2)\cdots f(\lambda_n)$$

and

$$\mathrm{Trace}[f(A)] = \mathrm{Trace}[f(D)] = f(\lambda_1) + f(\lambda_2) + \cdots + f(\lambda_n)$$

If we now assume that $f(A) = e^{At}$ in the above formulas, we have

$$\begin{aligned}
\mathrm{Det}(e^{At}) &= e^{\lambda_1 t}e^{\lambda_2 t}\cdots e^{\lambda_n t} \\
&= e^{(\lambda_1 + \lambda_2 + \cdots \lambda_n)t} \\
&= e^{(\mathrm{Trace}\ D)t} \\
&= e^{(\mathrm{Trace}\ A)t}
\end{aligned} \tag{3.14}$$

Returning to (3.13), we can again argue that a volume V of perturbations in phase space will be conserved if $|\mathrm{Det}(e^{At})| = 1$ [or taking into account Eq. (3.14) if $|e^{(\mathrm{Trace}\ A)t}| = 1$ or if Trace $A = 0$]. Thus, a flow represents a conservative system if the trace of the Jacobian is zero and a dissipative system if $|e^{(\mathrm{Trace}\ A)t}| < 1$, which translates to Trace $A < 0$. Since Trace $A = \lambda_1 + \lambda_2 + \cdots + \lambda_n$, the sum of the eigenvalues dictates whether volumes contract or expand or remain the same. Each eigenvalue gives the rate of contraction or expansion along a direction of one of the coordinates of the phase space. If all λ's are negative, the volumes contract along all directions. Obviously, we can have positive and negative λ's while $\lambda_1 + \lambda_2 + \cdots + \lambda_n < 0$. Thus, it is possible to have expansion along certain directions even though the initial volume shrinks in time. Such systems will be later termed chaotic systems. The eigenvalues $\lambda_1, \lambda_2, \ldots,$ λ_n are known as the Lyapunov exponents of the flow.[46] Direct extension of the above arguments to maps where $\mathbf{x}'_{n+1} = A\mathbf{x}'_n$ leads to the conclusion that the Lyapunov exponents are the logarithms of the eigenvalues of A. Note that in dissipative systems even though an initial volume shrinks to zero it does not necessarily mean that the volume will shrink to a point. In a 3D phase space a surface has zero volume, but it is not a point.

Examples

 a. The Lorenz system

$$\dot{x} = -\sigma x + \sigma y$$
$$\dot{y} = -xz + rx - y$$
$$\dot{z} = xy - bz$$

where σ, r, b are positive constants, is a dissipative system since

$$A = \begin{pmatrix} -\sigma & \sigma & 0 \\ -z + r & -1 & -x \\ y & x & -b \end{pmatrix}$$

and Trace $A = -(\sigma + 1 + b) < 0$.

 b. The Rössler system

$$\dot{x} = -y - z$$
$$\dot{y} = x + ay$$
$$\dot{z} = bx - cz + xz$$

where a, b, c are positive constants, has a Jacobian equal to

$$A = \begin{pmatrix} -0 & -1 & -1 \\ 1 & a & 0 \\ b & 0 & -(c+x) \end{pmatrix}$$

The trace of A is equal to $a - c - x$. The equilibrium points are the points $(0, 0, 0)$ and $(ab + c, (ab + c)/a, -(ab + c)/a)$. Thus, around the equilibrium point $(0, 0, 0)$ (i.e., $x \approx 0$) Trace $A = a - c$ and the system is dissipative when $a < c$. Around the other equilibrium point the flow is dissipative if $b < -1$.

   c. *The Hénon map*

$$x_{t+1} = 1 + y_t - ax_t^2$$

$$y_{t+1} = bx_t$$

corresponds to a dissipative system if

$$|\text{Det}A| = |b| < 1$$

2. INTEGRABLE AND NONINTEGRABLE DYNAMICAL SYSTEMS

   Consider the Hamiltonian system

$$\ddot{x} = -x - 2xy$$
$$\ddot{y} = -y - y^2 - x^2 \tag{3.15}$$

This is a system of two nonlinear coupled differential equations. If we set $z = \dot{x}$ and $w = \dot{y}$, we can write this system as a set of four first-order ODEs. We then find that the trace of the Jacobian is zero and that the system is conservative.

   In the formulation of system (3.15) two variables are involved. Thus, the number of degrees of freedom, N, is 2. We now change the variables from x and y to m and n, where $m = x - y$ and $n = x + y$. Thus, $x = (m + n)/2$ and $y = (n - m)/2$. If we substitute x and y in system (3.15), we obtain

$$\ddot{m} + \ddot{n} = -m - n + m^2 - n^2$$
$$\ddot{n} - \ddot{m} = -n + m - n^2 - m^2$$

By first adding and then subtracting these equations, we arrive at

$$\ddot{n} = -n - n^2$$
$$\ddot{m} = -m + m^2$$

Thus, with the change in variables we end up with a system of two nonlinear differential equations that are not coupled (they are independent). Each one is a function of one variable. We effectively are left with two $N = 1$ conservative systems with periodic evolutions. When this is possible, we call the $N = 2$ system integrable. Our solar system may be thought of as an integrable system, since the motion of each planet is "decoupled" from the motion of any other planet. Consequently, the study of planetary orbits does not involve interactions between the planets, which would make such a study very complicated.

If instead we consider the conservative system

$$\ddot{x} = -x - 2xy$$
$$\ddot{y} = -y + y^2 - x^2$$

(3.16)

we find that it cannot be separated into two independent $N = 1$ systems, and thus it is nonintegrable. [Eq. (3.16) is the famous Hénon–Heiles system, which describes the motion of a star in the effective potential due to the other stars in a galaxy, as well as other phenomena in physics and chemistry.] If we think of system (3.15) as a 2-degree-of-freedom system consisting of two particles, we can say that integrability allows us to speak about two independent particles rather than two interacting particles. Integrability eliminates their interaction, thus basically eliminating the potential energy. In spite of the presence of nonlinear terms, integrable systems thus exhibit smooth regions in phase space. In contrast, nonintegrable systems can exhibit very complicated "random-looking" motion. We come back to this in greater detail in Chapter 6, where we show that all dissipative systems are nonintegrable. As a corollary, all Hamiltonian (conservative system with $N = 1$ are completely integrable.

3. ERGODIC VERSUS NONERGODIC DYNAMICAL SYSTEMS

Because a conservative system cannot "forget" its initial state or prep-
aration, conservative systems cannot explain the approach to thermody-
namic equilibrium. To provide an explanation for the approach to ther-
modynamic equilibrium, the *ergodic hypothesis* was introduced. According
to this hypothesis, a system is called ergodic if a long time average of the
system is equal to ensemble averages of many different systems (but having
the same Hamiltonian). An ensemble is a great number of systems of the
same nature differing in configurations and velocities at a given instance,
such that they embrace every conceivable combination of configurations
and velocities. Since a very large number of such different systems would
uniformly cover the energy surface, the ergodic hypothesis is true if a single
trajectory would cover the energy surface uniformly. A simple example of
an ergodic system is given by the motion on a two-dimensional unit square
obeying the equations

$$\frac{dx}{dt} = 1 \qquad \frac{dy}{dt} = \mathbf{a} \tag{3.17}$$

The solution of this system is

$$x(t) = x_0 + t$$
$$y(t) = y_0 + \mathbf{a}t \tag{3.18}$$

From Eqs. (3.18) we can deduce an expression for the trajectory in phase
space (x, y):

$$y = y_0 + \mathbf{a}(x - x_0)$$

In our case the Hamiltonian of system (3.17) is $H = y - \mathbf{a}x = y_0 - \mathbf{a}x_0$
= constant. Thus, the system is conservative; hence, the energy surface is
the whole square.

Following the proof in the insertion and taking $\omega_1 = \mathbf{a}$ and $\omega_2 = 1$,
we see that if $\mathbf{a}$ is rational (i.e., $\mathbf{a} = m/n$, where m and n are integers),
then the trajectory is periodic, and if $\mathbf{a}$ is irrational the trajectory fills
the surface of the square. When the trajectory is periodic, it does not
fill the surface and therefore the system is not ergodic. When $\mathbf{a}$ is irra-

tional, the trajectory satisfies the ergodic hypothesis. We will later see that for dissipative systems the energy surface must shrink to some region in the phase space of zero volume. As a result, dissipative systems are, in general, ergodic.

A function $f(t)$ is periodic when $f(t) = f(t + T)$ for all t. Therefore, if $f(t)$ is the sum of two periodic functions

$$f(t) = \cos\omega_1 t + \cos\omega_2 t$$

it is periodic if

$$\cos\omega_1 t + \cos\omega_2 t = \cos\omega_2(t + T) + \cos\omega_2(t + T)$$

Since

$$\cos\omega(t + T) = \cos\omega t \cos\omega T + \sin\omega t \sin\omega T$$

it follows that it is possible to find an integer n so that $\omega T = 2\pi n$. Then $\cos\omega(t + T) = \cos\omega t$. Therefore, $f(t)$ is periodic if $\omega_1 t = 2\pi n$ and $\omega_2 t = 2\pi m$, or if $\omega_1/\omega_2 = n/m$, a ratio of two integers (or, in more general form, if there exist two nonvanishing integers k_1, k_2 such that $k_1\omega_1 + k_2\omega_2 = 0$). If ω_1/ω_2 is an irrational number (i.e., ω_1/ω_2 is not a ratio of two integer numbers), then the motion is called quasi-periodic and is represented by a trajectory that moves on a torus slowly filling its surface and without ever closing itself.

4. AUTONOMOUS SYSTEMS

When the formulation of a dynamical system is of the form

$$\dot{x}_1 = f_1(x_1, x_2, \ldots, x_n)$$
$$\dot{x}_2 = f_2(x_1, x_2, \ldots, x_n)$$
$$\vdots$$
$$\dot{x}_n = f_n(x_1, x_2, \ldots, x_n)$$

$$(3.19)$$

where $f_1, f_2, \ldots, f_n$ are functions of the observables $x_1, x_2, \ldots, x_n$ only (i.e., they do not depend explicitly on time), then we say that there are no time-varying forces acting on the dynamical system from the outside and that the vector field **f** is stationary. Such a system is called *autonomous*. The phase space of an autonomous system is a coordinate system with coordinates $x_1, x_2, \ldots, x_n$.

Since the right side of system (3.19) is stationary, it can be proven that no two trajectories (corresponding to two evolutions from two different initial conditions) cross through the same point in phase space.[36,175] If they did cross, that point would correspond to a single state, and from a single state we cannot have two different evolutions. Keep in mind that this is not the same as two trajectories approaching an equilibrium point as $t \rightarrow \infty$, which is allowed.

If, on the other hand, the formulation of a dynamical system is of the form

$$\dot{x}_1 = f_1(x_1, x_2, \ldots, x_n, t)$$
$$\dot{x}_2 = f_2(x_1, x_2, \ldots, x_n, t)$$
$$\vdots$$
$$\dot{x}_n = f_n(x_1, x_2, \ldots, x_n, t)$$

(3.20)

where now $f_1, f_2, \ldots, f_n$ depend explicitly on time, the system is called *nonautonomous*. In this case, the right side of system (3.20) is nonstationary, and thus any two trajectories originating from two different initial conditions may cross through the same point in phase space. To construct a phase space where the noncrossing trajectory principle holds, we must augment the phase space by one dimension so that the coordinates are $x_1, x_2, \ldots, x_n, t$:

$$\dot{x}_1 = f_1(x_1, x_2, \ldots, x_n, t)$$
$$\vdots$$
$$\dot{x}_n = f_n(x_1, x_2, \ldots, x_n, t)$$
$$\dot{t} = 1$$

In this way, time becomes one of the observables, and the right side effectively becomes stationary; therefore, in this phase space the noncrossing trajectory property holds.

Example. The nonautonomous system

$$\ddot{x} + k\dot{x} + f(x) = F(t)$$

where $F(t)$ represents some external forcing, can be transformed to an autonomous system if we rewrite it as

$$\dot{x}_1 = x_2$$
$$\dot{x}_2 = -kx_2 - f(x_1) + F(t)$$
$$\dot{t} = 1$$

CHAPTER 4

ON FRACTALS

1. TOPOLOGICAL, EUCLIDEAN, AND FRACTAL DIMENSIONS

Ever since fractal geometry was introduced,[140] it has become an indivisible part of the theory of dynamical systems and chaos. Many good books have been written on fractal geometry,[57,141,163,164] so we will avoid a lengthy discussion of it. Instead we concentrate on some basic definitions and ideas that we will recall throughout this book. We start with a discussion of the notion of dimension.

Let us assume that we want to completely characterize a tree. To achieve that, we have to measure its weight, height, width, density, color, etc. In fact, we have to obtain as many measurements as necessary to distinguish this tree from anything else. Only after we obtain them can we completely characterize the tree. We need all these data because there is no master characteristic with absolute discriminating power. We may be able to characterize a tree by its height only, but many other objects may have the same height.

We can extend these arguments to a set of points or any mathematical–geometrical object. To completely describe this object, we have to have many different measurements, called dimensions (the word *dimension* is derived from the Latin word *dimensio,* which means *measure*).

The most familiar dimension is the Euclidean dimension, followed in popularity by the topological dimension. Figure 13 introduces the reader to the difference between the definitions of these two dimensions.

Let us now consider a line segment of length L that can be divided into N parts of length l. We say that each part is scaled down by a ratio of $m_1 = 1/N$. Magnification of any of these parts by a factor of L/l will

49

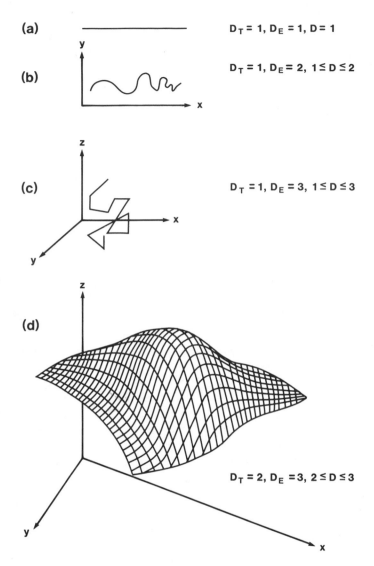

(a) ——————————————— $D_T = 1, D_E = 1, D = 1$

(b) $D_T = 1, D_E = 2, 1 \leq D \leq 2$

(c) $D_T = 1, D_E = 3, 1 \leq D \leq 3$

(d) $D_T = 2, D_E = 3, 2 \leq D \leq 3$

FIGURE 13. There are two basic definitions of dimension: the Euclidean (D_E) and the topological (D_T). They both can assume only the integer values 0, 1, 2, 3, but, for a specific object, they may be the same. To divide space, cuts that are called surfaces are necessary. Similarly, to divide surfaces, curves are necessary. A point cannot be divided since it is not a continuum. Topology tells us that, since curves can be divided by points that are not continua, they are continua of dimension 1. Similarly, surfaces are continua of dimension 2, and space is a continuum of dimension 3. Apparently, the topological dimension of a point is zero. According to the Euclidean definition, a configuration is called one dimensional if it is embedded on a straight line, two dimensional if it is embedded on a plane, and three dimensional if it is embedded in space. Therefore the Euclidean dimension of a straight line is 1, of curve (b) is 2, of curve (c) is 3, and of surface (d) is 3 (Tsonis and Tsonis[217]).

reproduce the whole original segment (actually after some appropriate translation it will be mapped exactly onto itself. We say that the line is a self-similar structure, since a small part, when magnified, can reproduce exactly a larger portion. This property is a kind of symmetry that is called *scale invariance* or, simply, *scaling*.

We can extend this example by considering a two-dimensional structure, such as a square area in the plane of side L, which can be divided into N self-similar parts of side l. Now each part is scaled in relation to the original square by a factor of $m_2 = 1/\sqrt{N}$. Magnification of any of the parts by a factor of L/l will reproduce the original structure exactly. Similarly, a three-dimensional object such as a solid cube of side L can be divided into N smaller cubes that are scaled down by a ratio of $m_3 = 1/N^{1/3}$. Now, we have $m_1 = 1/N$, $m_2^2 = 1/N$, and $m_3^3 = 1/N$, or $N = 1/m^D$, where $D = 1$ for the straight line, $D = 2$ for the square, and $D = 3$ for the cube. These values are simply the Euclidean dimensions of a straight line, a square, and a cube. We can now define the similarity dimension as[141]

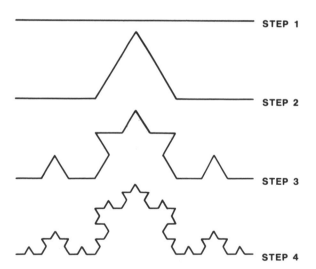

FIGURE 14.   The first four steps in constructing the Koch curve. At each step the middle third of every straight-line segment of length L is replaced by two equal segments of side $L/3$, forming an equilateral triangle. Proceeding for an infinite number of steps, one obtains the Koch curve.

$$D = \frac{\log N}{\log(1/m)} \qquad (4.1)$$

where N is the number of the parts that are scaled down by a ratio m from the original self-similar structure.

We now proceed with a somewhat puzzling example (see Fig. 14). As our first step, we consider a straight-line segment of length L. In step 2, we replace the middle third of this segment of two equal segments of side $L/2$ forming part of an equilateral triangle. By repeating this procedure many times, we can produce the famous Koch curve.[141]

In this figure, we may think of it as follows: At each step we produce a picture that is a part of the picture of the next step with a change of scale. Therefore, the curve is self-similar. For each scale change by three we need four parts. Thus, according to Eq. (4.1), $D = \log 4/\log 3 = 1.26$. Surprisingly, the similarity dimension is not an integer and is definitely less than its Euclidean dimension. Structures that possess a similarity dimension that is not an integer are called *fractals,* and the dimension is the *fractal dimension.*[141]

2. A GENERAL WAY TO MEASURE THE FRACTAL DIMENSION

From the foregoing arguments we have $m = l/L$. For some maximum $L = L_{max}$, it follows that

$$N = \left(\frac{l}{L_{max}}\right)^{-D}$$

or that

$$N(l) \propto l^{-D}$$

Thus, the number of scale-size l parts is proportional to the length l raised to the power $-D$, where D is the fractal dimension of the object. This leads to a general way of calculating D for an object embedded in n-dimensional Euclidean space: Cover the object with an n-dimensional grid of size l_1

and count the number $N(l_1)$ of the boxes that contain a piece of the object. Repeat for various sizes $l_2, l_3, \ldots, l_n$. Plot the logarithm of $N(l_1)$, $N(l_2)$, $\ldots, N(l_n)$ versus the logarithm of $l_1, l_2, \ldots, l_n$. If the resulting curve has a linear part over a wide range of scales, then the slope of that linear part is an estimate of the fractal dimension. This technique is called the box-counting technique (Fig. 15).

3. SOME INTERESTING PROPERTIES OF FRACTAL OBJECTS

The existence of noninteger similarity dimensions reflects some unusual properties of fractals. In our example with the construction of the Koch curve, one can calculate that at each step the length of the curve increases by a factor of $\frac{4}{3}$. If instead of one segment of side L we started with an equilateral triangle of side L and repeated the previous procedure for each side, then the perimeter of the boundary will increase at every step by a factor of $\frac{4}{3}$. Eventually, we end up with a boundary of infinite length enclosing a finite area (Fig. 16). Such mathematical curiosities cannot be found in Euclidean structures, but they are mathematical realities.

Consider the straight-line segment again. If one measures the length L of the segment with a yardstick of size l, one finds that $\log L(l)$ is a linear function of log with slope of zero. Thus, $L(l) \propto l^{1-D} [L(l) = N(l)l \propto l^{-D}l \propto l^{1-D}]$, where $D = 1$ is the similarity dimension of a straight line. This simply means that $L(l) = $ constant no matter what l is. If instead we

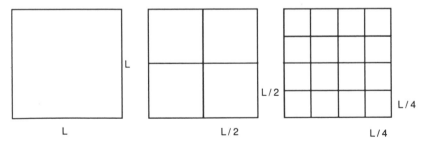

FIGURE 15. The idea behind the box-counting technique for estimating the fractal dimension. At first, we cover the object with a square of length $l_1 = L$. Then we divide the square into squares of length $l_1 = L/2$ and count the number of those squares, $N(l_1)$, that contain a piece of the object. Then we divide each square of length $L/2$ into squares of length $l_3 = L/4$ and repeat the procedure. In a $\log N(l)$ versus $\log l$ plot, the slope is an estimate of the fractal dimension. Note that to be statistically correct, the whole procedure should be repeated several times for different initial placings of the square of length L.

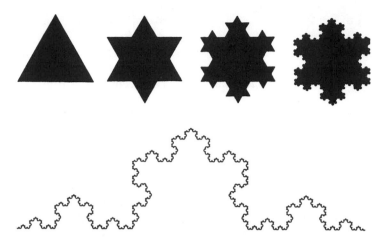

FIGURE 16.   If we repeat the procedure outlined in Fig. 14 but for every side of an equilateral triangle, we end up with a boundary of infinite length enclosing a finite area (this finite area is a little smaller than the area of the circle enclosing the initial equilateral triangle). (From Mandelbrot[141] reproduced by permission.)

consider the Koch curve and repeat the previous procedure, we find that $L(l) \propto l^{1-D}$ with $D = 1.26$, which is the similarity dimension of the Koch curve. Since $D > 1$, then, as $l \to 0$, $L(l) \to \infty$, in agreement with the preceding discussion. Thus, when we deal with a fractal boundary such as the Koch curve, the length of some part of this curve will be different for different yardstick lengths.

If we think of the yardstick length as a spatial resolution, it is obvious that the measured properties of a fractal are resolution dependent! It turns out that such a relation can be found for structures that are not exactly self-similar, such as a coastline,[141] where a small piece when magnified looks like, but it is not *exactly* the same as, a piece at a larger scale. The very existence of a linear relationship between $\log L(l)$ and $\log l$ [or $\log N(l)$ and $\log l$] and, thus, the existence of some D, indicates that the small scales are related to the large scales via a fractal dimension. This type of scaling is called statistical self-similarity and reflects the fact that a structure may look statistically similar while being different in detail at different scales (see, for example, Tsonis and Elsner[209]).

Apart from the above-defined fractal dimension, we can define many other relevant dimensions that characterize a set of points. Details on all these multiple fractal dimensions are presented in Section 3 of Chapter 5.

4. "FRACTALITY" IN TIME SERIES

When an object is fractal, the scaling dictates that the small scales are related to the large scales. This relationship is quantified by the fractal dimension. The relationship between small and large scales translates to having a large part equivalent (in a statistical sense) to a small part magnified by some factor. In other words, when a small scale is magnified, it reveals details similar to those observed over larger scales (see Fig. 17). Granted that such self-similarity may describe many objects in nature, is it possible that it also applies to time sequences? Let us assume that the answer to this question is yes. Let us also assume that we are examining the scaling properties of a time series representing wind speed at a point. Zooming on a small time scale (say 1 min), we may discover that the wind fluctuates over that time scale between 1.0 and 2.0 m/sec. If the record were self-similar when this small scale is magnified by a factor of 24, it should reproduce the fluctuation over a 24-h time scale. Thus, if the record is self-similar, the fluctuations over a 24-h interval should range from 24 × 60 m/sec to 24 × 60 × 2 m/sec. That is definitely not true in observables from nature. Therefore, the type of self-similarity discussed for spatial objects cannot apply to time records. Nevertheless, many natural phenomena are described in the space–time domain and not just in space. Is it possible that some other kind or some modified version of self-similarity exists in time records?

A random walk is a process according to which a particle moves from some initial position in a random direction with a step length that is a random variable having some predescribed probability distribution. Because this is the motion observed of a "Brownian particle," random walks are often referred to as Brownian motion. In one dimension starting from position 0, the random walker may take a "step" to the right or left. Each direction has a probability 0.5. If we assume that the length of each step is a random variable w having the $N(0, 1)$ distribution, then an obvious way to generate a record of the position in time (or a trace of a Brownian motion) is given by

$$x(t) = \sum_{i=1}^{t} w_i$$

Such a record (trace) is shown in Fig. 18.

We can extend this record to higher dimensions. For example, in three dimensions

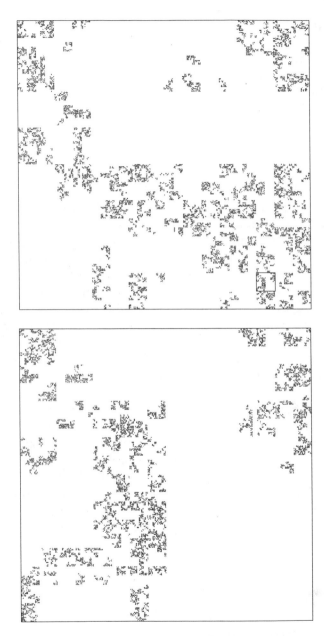

FIGURE 17.   A statistically self-similar fractal (top). The bottom figure is a magnification of the small square in the top figure. The magnified part does not reproduce the whole part exactly, but it reveals similar details to those observed over larger scales. Both have fractal dimension of about 1.7.

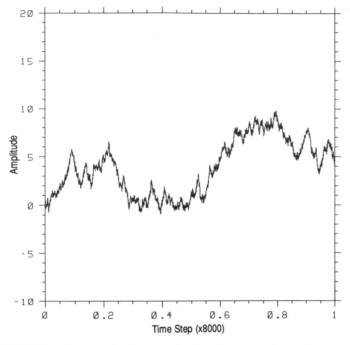

FIGURE 18.   An example of a *trace* of a Brownian motion is one dimension.

$$x(t) = \sum_{i=1}^{t} w_i$$

$$y(t) = \sum_{i=1}^{t} w_i'$$

$$z(t) = \sum_{i=1}^{t} w_i'' \qquad (4.2)$$

In a three-dimensional Cartesian coordinate system, system (4.2) defines a *trail* for that Brownian motion (see Fig. 19). This trail is a self-similar (fractal) path. In fact, all trails of Brownian motion are self-similar.[141] Nevertheless, the record in Fig. 18 is not self-similar: It is known that for a Brownian motion in n dimensions and for a large number of walkers, the average displacement is a random number x with mean $\bar{x} = 0$ and standard deviation $s = \sqrt{2nt} \propto t^{0.5} = t^H$. Since $s \propto t^{0.5}$, the departure from the origin increases with time, and therefore the position of a walker at

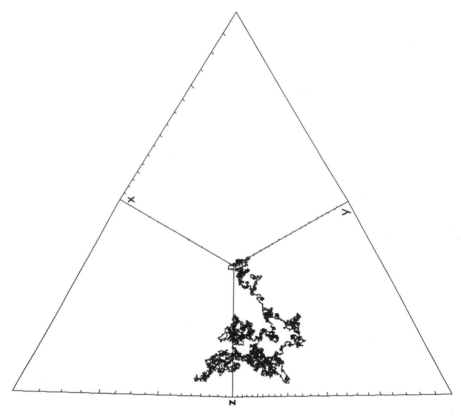

FIGURE 19. A *trail* of a Brownian motion in three dimensions after 16,000 steps. The trail is a fractal path.

some time is correlated to the position at an earlier time. Even though the walker tends to "explore" a given region rather thoroughly (because explorations over short distances can be done in much shorter times than exploration over long distances), he or she actually "wanders" away, slowly increasing his or her departure from the origin. Thus, in Brownian motion it is the displacement at a given time, not the actual position of the walker, that is independent of the past. With such considerations it follows (see also Feder[57]) that for any $n = 1$ motion,

$$\Delta x = x(t_1) - x(t_0) \propto w(t_1 - t_0)^H = \Delta t^H \qquad (4.3)$$

From Eq. (4.3) it follows that $E(\Delta x) = 0$ and that $\text{Var}(\Delta x) = E[(\Delta x)^2] \propto \Delta t^{2H}$.

Also, according to Eq. (4.3), if we magnify Δt by a factor λ (it becomes $\lambda \Delta t$), then $(\lambda \Delta t)^H = \lambda^H \Delta t^H = \lambda^H \Delta x$. Thus, Δx must be magnified by a factor λ^H. As a consequence, a magnified small scale could be statistically equivalent to large scales if the magnification factors are different for different coordinates. This nonuniform scaling where shapes are statically invariant under transformations that scale different coordinates by different amounts is called *self-affinity*. A formal definition of a self-affine signal is as follows: We say that $x(t)$ is self-affine if for some arbitrary Δt the following relation holds: [141]

$$\Delta x(\lambda \Delta t) \overset{d}{=} \lambda^H \Delta x(\Delta t) \qquad (4.4)$$

for all $\lambda > 0$. The symbol $\overset{d}{=}$ denotes identity in statistical distributions, and $0 < H < 1$ is the scaling exponent. Equation (4.4) indicates that the distribution of increments of x over some time scale $\lambda \Delta t$ is identical to the distribution of the increments of x over a lag equal to Δt multiplied by λ^H. Therefore, if time is magnified by a factor λ, then x is magnified by a factor λ^H.

In Eq. (4.4), H is allowed to range between 0 and 1. Pure Brownian motion corresponds to $H = \frac{1}{2}$. For any other value the motion is called fractional Brownian motion (fBm). The restriction in the range of H is due to probability considerations. [141] A consequence of the foregoing discussion is that the trace of an fBm, $x(t)$, is a nonstationary process. The degree of nonstationary depends on the value of H. For $H = 0$ the position becomes independent of time (recall that $s \propto t^H$), and, therefore, the position at a given instant is independent of the position at any time in the past. The process is clearly stationary and corresponds to white noise (Fig. 2). As H increases, the dependence of the position at a given instant to the position in the past also increases. In addition, the departure from the origin increases. Thus, in effect the exploration of a given area decreases, and the walker leaves the area sooner. As a result, the trail (or the trace) becomes smoother (see Fig. 20).

The concept of fractal dimension is strongly connected to self-similar (uniform) scaling. Thus, extending it to self-affine shapes can be tricky. [141,142] As is clearly explained in Peitgen and Saupe, [164] the fractal dimension of an fBm can be 1, $1/H$, or $2 - H$, depending on the measurement technique and choice of length scale! *The fractal dimension, however, of the trail of a Brownian motion in an n-dimensional Euclidean space is $1/H$ and independent of n.* A qualitative proof goes as follows: for

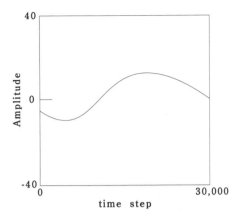

FIGURE 20.   The trace of a fractional Brownian motion with H = 3.5. Note that, compared to Fig. 18 (which corresponds to H = 0.5), this trace is smoother.

$H \rightarrow 1$ the traces are very smooth, and, thus, in any high-dimensional space where every coordinate is tracing out an independent fBm, the resulting trail is smooth ($D \rightarrow 1$). If $H \rightarrow 0$, the traces are more like white noise, and they simply fill the embedding space ($D \rightarrow \infty$). Thus, in general, $D = 1/H$.

5. POWER SPECTRA FOR FRACTIONAL BROWNIAN MOTIONS AND THE EXPONENT H

If a random function $x(t)$ exhibits power spectra with equal power at all frequencies, it is called white noise (in analogy to white light, which is made up of waves of equally distributed frequencies in the visible range of electromagnetic radiation). In such a case $S(f)$ = constant or $S(f)$ $\propto 1/f^0$. For $S(f) \propto 1/f^2$ we obtain the brown noise that corresponds to a trace of a Brownian motion. In general, a time series $x(t)$ with a spectral density $S(f) \propto 1/f^a$ corresponds to a trace of an fBm with $H = (a-1)/2$. The proof goes as follows: Recall that if $x(t)$ denotes the trace of fBm with $0 < H < 1$, then the function

$$y(t) = \frac{1}{\lambda^H} x(\lambda t) \qquad (4.5)$$

for $\lambda > 0$ has the same probability distribution as $x(t)$. Thus, it has the same spectral density. If we assume that the processes are defined in time from 0 to some length T, then we may write

$$F_y(t) = \int_0^T y(t)e^{-2\pi ift}\, dt \tag{4.6}$$

as the Fourier transform of $y(t)$.

Taking into account Eqs. (4.5) and (4.6), we have

$$F_y(f) = \frac{1}{\lambda^H} \int_0^{\lambda T} x(v)e^{-2\pi i(f/\lambda)v}\, dv$$

where we set $v = \lambda t$. Then

$$F_y(f) = \frac{1}{\lambda^{H+1}} F_x\left(\frac{f}{\lambda}\right) \tag{4.7}$$

Now recall from Chapter 2 that

$$S(f) = \frac{1}{T} |F(f)|^2$$

Thus, Eq. (4.7) becomes

$$S_y(f) = \frac{1}{\lambda^{2H+1}} \frac{1}{\lambda T} \left|F_x\left(\frac{f}{\lambda}\right)\right|^2$$

As $\lambda T \to \infty$, we conclude that

$$S_y(f) = \frac{1}{\lambda^{2H+1}} S_x\left(\frac{f}{\lambda}\right)$$

Remember now that y is a properly rescaled version of x. Thus, the spectral densities of x and y must coincide. Therefore, $S_y(f) = S_x(f)$ and

$$S_x(f) = \frac{1}{\lambda^{2H+1}} S_x\left(\frac{f}{\lambda}\right)$$

If we now formally replace λ with f we arrive at

$$S_x(f) \propto \frac{1}{f^{2H+1}} = \frac{1}{f^a} \tag{4.8}$$

Figure 21 shows the power spectra of the trace in Fig. 18. This is a $\log S(f)$ versus $\log f$ plot. Over a wide range of frequencies the slope of the plot is around 2, as expected when $H = 0.5$.

How To Generate an fBM Trace with a Desired Spectral Density Function. A time series x_i, $i = 1, \ldots, N$, whose spectral density function satisfies the relation $S(f) = Cf^{-a}$, where $a = 2H + 1$, is obtained via the relation

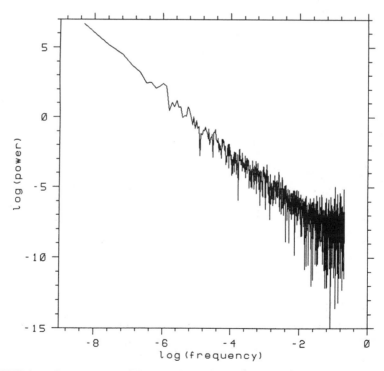

FIGURE 21.   Power spectra of the Brownian motion of Fig. 18. The spectral density $S(f)$ scales with frequency f according to $S(f) \propto f^{-a}$ with $a \approx 2.0$.

$$x_i = \sum_{k=1}^{N/2} \left[Ck^{-a} \left(\frac{2\pi}{N} \right)^{1-a} \right]^{1/2} \cos \left(\frac{2\pi ik}{N} + \phi_k \right)$$

where C is a constant and ϕ_k are $N/2$ random phases randomly distributed in $[0, 2\pi]$.

PART II

THEORY

Since nothing accidental is prior to the essential neither are accidental causes prior. If, then, luck or spontaneity is a cause of the material universe, reason and nature are causes before it.

—ARISTOTLE, METAPHYSICS, BOOK XI, CH.8

CHAPTER 5

ATTRACTORS

Let us again consider the motion of a pendulum. As we saw in the example in Chapter 2, the coordinates of the phase space are x_1 and x_2, where x_1 stands for the position and x_2 stands for the velocity. The position x_1 is measured by the angle of the pendulum for the vertical (see also Fig. 10). You may recall that there exist two equilibrium states $\bar{x}_1 = 0$, $\bar{x}_2 = 0$ and $\bar{x}_1 = 0$, $\bar{x}_2 = \pi$. Starting the motion of the pendulum from some initial condition $x_1(0)$, $x_2(0)$, the mathematical possibilities are that the system may be (1) "attracted" by the stable equilibrium state, (2) repelled by the unstable equilibrium state, or (3) engaged in a never-ending motion around the stable equilibrium state. For all those possibilities in the real world only the first possibility is observed. We may take a system to some initial state close or far from the equilibrium state(s), but, due to friction, after some time the system settles down to the equilibrium state that is observable (in our example the fixed point $\bar{x}_1 = 0$, $\bar{x}_2 = 0$). If we remove friction, the pendulum moves in an endless periodic motion, with the period being defined by the initial condition. The settling-in part of the evolution (or the transient) is modeled by a *trajectory* in the phase space (when we are dealing with flows). A phase portrait is a graph depicting the evolution of the system from different initial conditions (see Fig. 22). The final state or the equilibrium state is modeled by *limit sets*.

We have already "met" (Chapter 2) four such limit sets, all of them points: fixed point, repeller, center, and saddle. From those limit sets only fixed points collect trajectories. By definition a limit set that collects trajectories is called an *attractor*. Thus, from the limit points only the fixed points are attractors (point attractors). Without friction the pendulum is simply a conservative system whose motion is undistributed, and no energy is needed to be spent. When friction is introduced in order to overcome

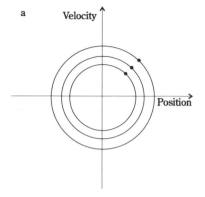

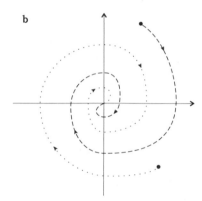

FIGURE 22.   Phase portrait for (a) frictionless pendulum and (b) pendulum with friction. The dots represent initial conditions.

its opposition, energy must be spent. Thus, the kinetic energy is gradually being spent, and eventually the pendulum stops at the stable equilibrium point. The pendulum with friction is obviously a dissipative system. *The point* is that *conservative systems never settle to an equilibrium state and, therefore, do not exhibit attractors.*

Points, however, are not the only limit sets. A cycle may also be a limit set for a trajectory. An example of such an attractor can be given by the swinging pendulum of the grandfather clock, where the effect of friction is compensated via a mainspring. We may disturb the regular periodic motion of the pendulum, but soon the pendulum assumes its regular periodic motion, whose frequency is dictated by the specification of the mainspring. Another familiar system that exhibits a limit cycle (periodic) final state is our heart. Our heartbeat is regular and periodic. We may

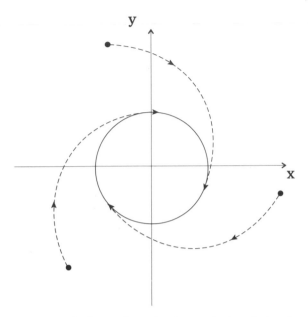

FIGURE 23. Phase portrait of a two-dimensional system having a limit cycle as an attractor. Trajectories from different initial conditions are attracted and stay on the cycle. The evolution of the system is periodic.

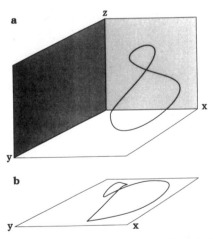

FIGURE 24. (a) A periodic trajectory in 3D; (b) its projection on a plane.

disturb this cycle, or we may take the "system" to various initial conditions (due to panic, joy, exercise, etc.), but soon after the "disturbance" the beat assumes its regular cycle (Fig. 23). Such periodic motions should not be confused with periodic motions like that of the frictionless pendulum that are not associated with attractors. A periodic attractor of a limit cycle may be embedded in more than two dimensions. In such cases the trajectory might assume all sorts of interesting shapes, which, when projected on a plane, might give the impression that the trajectory intersects itself, which is not allowed (Fig. 24).

Having discussed points and limit cycles, we now consider the following scenario. Assume that we have two identical but uncoupled frictionless pendulums of unit mass. What describes their combined motion? Each pendulum is an $N = 1$-degree-of-freedom conservative system, and thus each one is an integrable system. The formulation of their motion is given by the equations

$$\ddot{\theta}_1 + \frac{g}{l}\sin\theta_1 = 0$$

$$\ddot{\theta}_2 + \frac{g}{l}\sin\theta_2 = 0$$

where θ indicates the angle from the resting position. The phase space of each of them is two dimensional with coordinates θ and $\dot{\theta}$. As we already know, each motion is periodic, with the periodicity being determined

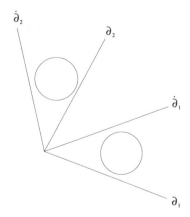

FIGURE 25.   Phase space of a system of two uncoupled frictionless pendulums. Each pendulum performs a periodic motion on its two-dimensional space with coordinates ∂ (position) and $\dot{\partial}$ (velocity). Thus, the dimensionality of the combined motion is four dimensional with coordinates ∂_1, ∂_2, $\dot{\partial}_1$, $\dot{\partial}_2$. However, due to the conservation of energy the effective dimensionality is 3.

by the initial conditions. Since the pendulums are uncoupled, the combined phase space is four dimensional with coordinates θ_1, θ_2, $\dot{\theta}_1$, $\dot{\theta}_2$ (see Fig. 25).

The phase-space trajectories are two closed orbits, one of which is confined on the $\theta_1\dot{\theta}_1$ plane and the other on the $\theta_2\dot{\theta}_2$ plane. The Hamiltonians are

$$H_1 = \frac{l^2\dot{\theta}_1}{2} - lg(1 - \cos\theta_1)$$

$$H_2 = \frac{l^2\dot{\theta}_2}{2} - lg(1 - \cos\theta_2)$$

Since the total energy is conserved (recall that we have assumed that each system is conservative), $H_1 + H_2$ = constant. Consequently,

$$l^2\frac{\dot{\theta}_1}{2} - lg(1 - \cos\theta_1) + l^2\frac{\dot{\theta}_2}{2} - lg(1 - \cos\theta_2) = \text{constant}$$

or

$$\dot{\theta}_1 = \frac{2g(1 - \cos\theta_1)}{l} + \frac{2g(1 - \cos\theta_2)}{l} - \dot{\theta}_2 + \frac{2 \cdot \text{constant}}{l^2}$$

Therefore, any one of the coordinates of the combined phase space can be expressed as a function of the remaining three. The "effective" phase space is thus three dimensional. Therefore, the combined motion takes place on an object embedded in a three-dimensional phase space whose projections are circles. Assuming that the independent periodic motions are of period $2\pi/\omega_1$ and $2\pi/\omega_2$, respectively, if $\omega_1/\omega_2 = n/m$ (where n and m are integers) then the combined motion is periodic and the object that describes the combined motion would be some cycle (recall that insertion in Chapter 3). Otherwise the motion would be quasi-periodic. Recall that in such motion the trajectory fills the surface of a torus. In this sense quasi-periodic motions can look quite "irregular." In this example, however, the cycle or the torus is not an attractor (as we expect when dealing with conservative systems), much like the periodic motion around a center for a single frictionless pendulum is not an attractor. The size of the torus depends on the combination of the initial conditions of the two uncoupled pendulums. In this case the phase portrait consists of toroidal surfaces nested inside each other (Fig. 26).

Quasi-periodic motions occur in nature quite often (ideally the daily temperature is a quasi-periodic variable with two distinct frequencies, $2\pi/$ 24 h and $2\pi/365$ days). Since we can hardly assume that nature is a conservative system, are there any circumstances under which a system may exhibit a torus as an attractor? Recall how a single frictionless pendulum is transformed to a dissipative system with a limit cycle attractor: We introduce friction, and then we compensate the effect of friction via a mainspring. If we do the same to both pendulums, we might end up with a system of two uncoupled oscillators whose combined behavior is dictated by an attracting invariant torus. In fact, according to the Peixoto theorem of mathematical dynamics, even if we somehow couple the two oscillators, their trajectories still might be attracted to an invariant torus.

Up to this point we have discussed three types of attractors: points, limit cycles, and tori. All of them are submanifolds of the total available phase space. In addition, they are topological structures characterized by topological dimensions of 0, 1, and 2, respectively. Note that the notion of torus can be extended to higher dimensions to obtain hypertori embedded in n dimensions. The identification of these attractors from observables is quite straightforward. Fourier analysis can verify if a given evolution is steady state, periodic, or quasi-periodic (see Chapter 2). A very interesting property of systems that exhibit such attractors is that *the long-term predictability of these systems is guaranteed.* When a system that exhibits a periodic attractor is "started" from many different initial conditions, it will always become periodic with a fixed period.

When a system has a torus as its attractor, a set of different initial

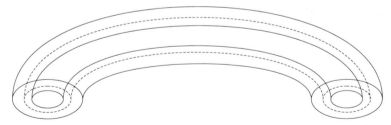

FIGURE 26.   The combined motions of the system shown in Fig. 25 take place on an object embedded in three dimensions whose projections are circles. Such an object is called a torus. However, since the example refers to conservative systems, the torus is not an attractor. Its size is determined by the initial conditions of the two pendulums. Thus, the phase portrait consists of toroidal surfaces nested inside each other.

conditions defines a set of different trajectories each one rotating around the torus, gradually filling its surface but without diverging as time goes on. They stay as close as they were to begin with. This also provides some kind of long-term predictability, since errors in the initial conditions would not grow (Fig. 27). The future value will always differ from its actual value by a constant amount. Once we observe this amount, the prediction becomes excellent. An explanation for this is provided later.

1. STRANGE ATTRACTORS

Figure 28 shows the spectra of some observable. Here we do not observe just one or several frequencies. The power is distributed over a range of frequencies, thus generating a broadband power spectrum. Such spectra are indicative of nonperiodic random evolutions where there is motion in all frequencies. Is it possible that such spectra arise as a result of a deterministic dynamical system? Let us assume that the answer to this question is yes. For this to be true we must have a nonperiodic trajectory in the phase space (or, strictly speaking, periodic with an infinite period) and which never crosses itself (recall the noncrossing property of trajectories).

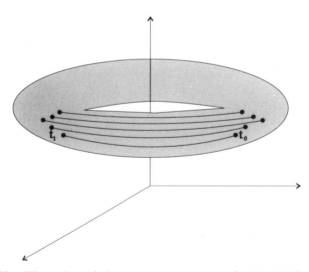

FIGURE 27.   When a dynamical system possesses a torus as its attractor, then a set of close initial conditions define a set of trajectories going around and around the torus, each one gradually filling its surface while not diverging.

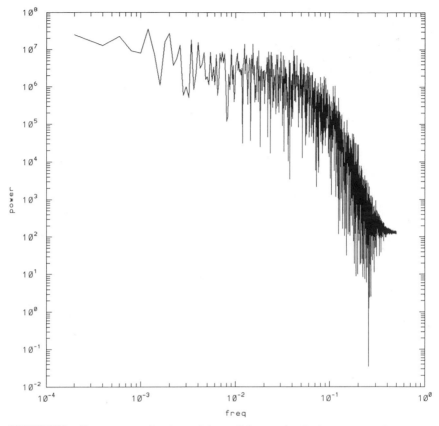

FIGURE 28. Power spectra showing activity at all frequencies. Such spectra are often called broadband noise spectra and are indicative of nonperiodic motion.

The trajectory must thus be of infinite length and confined in a finite area of the phase space of zero volume. Since the attractor cannot be a torus (then the trajectory is of infinite length and confined to a finite area), the only alternative is that the attracting set in question is a fractal set (recall the infinite-length boundary of the Koch curve enclosing a finite area).

The first such dynamical system was derived by Lorenz[135] from the convection equations of Saltzman.[188] This system, described by the following three differential equations, gives an approximate description of a horizontal fluid layer heated from below. The fluid at the bottom gets warmer and rises, creating convection. For a choice of the constants that

correspond to sufficient heating, the convection may take place in an irregular and turbulent manner:

$$\frac{dx}{dt} = ax + ay$$

$$\frac{dy}{dt} = -xz + bx - y$$

$$\frac{dz}{dt} = +xy - cz \tag{5.1}$$

where x is proportional to the intensity of the convection motion, y is
proportional to the horizontal temperature variation, z is proportional to
the vertical temperature variation, and a, b, and c are constants. Figure
29 depicts the path of a trajectory in the state space (x, y, z) for $a = 10$,
$b = 28$, $c = \frac{8}{3}$ after integrating system (5.1) for 10,000 time steps. The first
5000 points are colored blue, and the rest are green. The apparent white
and yellow parts are actually blue and green lines so close together that
the photographic device cannot distinguish them. Figures 29a–c help reveal
the two-lobed nonplanar shape and thickness of the attractor. Figure 29d
is a closeup of the orbit, which even for an infinite number of points does
not intersect or repeat itself. The blue and green lines interleave throughout
the attractor (a–c), with the interleaving existing on all scales (d).

Obviously, the Lorenz attractor does not look like the well-behaved
attractors previously described. The trajectory is deterministic since it is
the result of the solution of system (5.1), but is strictly nonperiodic. The
trajectory loops to the left and then to the right irregularly. Extensive studies
have shown that the fine structure of the Lorenz attractor is made up of
infinitely nested layers (infinite area) that occupy zero volume. One may
think of it as a Cantor-like set (see Fig. 30) in a higher dimension. Its
fractal dimension has been estimated to be about 2.06 (see, for example,
Grassberger and Procaccia[87]).

The fractal nature of an attractor does not merely imply nonperiodic
orbits; it also causes nearby trajectories to diverge. As with all attractors,
trajectories that are initiated from different initial conditions soon reach
the attracting set, but two nearby trajectories do not stay close to each
other (as with the torus). They soon diverge and follow totally different
paths in the attractor.

The divergence means that the evolution of the system from two

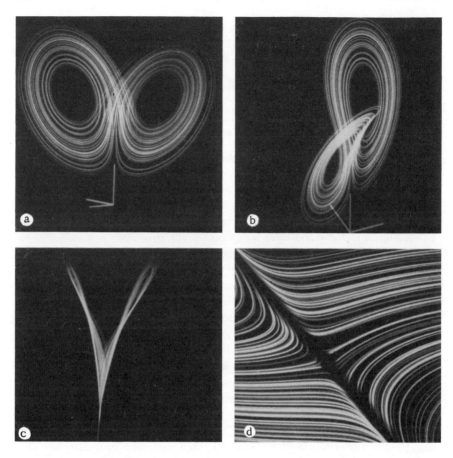

FIGURE 29.   (a)–(c). A trajectory of the Lorenz system with $a = 10$, $b = 28$, and $c = \frac{8}{3}$ as
seen from various angles. The trajectory is 10,000 time steps long. The first 5000 points are
plotted in blue and the rest in green. (d) A closeup of the trajectory. The blue and green lines
interleave throughout the attractor, with the interleaving existing on all scales. Apparent white
or yellowish parts indicate regions in the attractor where the trajectory returns very close to
a position it occupied before, but never crossing the path of its past. Because of the closeness,
the photographic device cannot distinguish between blue and green, thus producing white or
yellow colors. (Reproduced with permission from Dr. Gottfried Mayers-Kress, Los Alamos
National Laboratory. This figure appears in the book *From Cardinals to Chaos,* Cambridge
University Press, 1989.) For a color reproduction of this figure see the color plates beginning
facing page 148.

FIGURE 30.   The Cantor set begins with a line of length 1; then the middle third is removed, then the middle third of all the remaining intervals is removed, and so on. The Cantor set or Cantor "dust" is the number of points that remain. The total length of all intervals removed is $\frac{1}{3} + 2(\frac{1}{3})^2 + 4(\frac{1}{3})^3 + 8(\frac{1}{3})^4 + \cdots = 1$. Thus, the length remaining must be zero. Therefore, in the Cantor set the number of points is obviously infinite, but their total length is zero. The fractal dimension of this set is 0.6409. It is definitely greater than the topological dimension of a dust of points, which is zero.

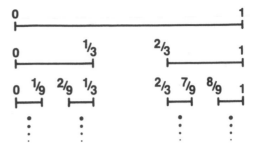

slightly different initial conditions will be completely different, as may be seen in Figs. 31a and b. The dot in Fig. 31a represents 10,000 initial conditions that are so close to each other in the attractor that they are visually indistinguishable. They may be viewed as 10,000 initial situations that differ only slightly from each other. If we allow these initial conditions to evolve according to the rules (equations) that describe the system, we see (Fig. 31b) that after some time the 10,000 dots can be anywhere in the attractor. In other words, the state of the system after some time can be anything despite the fact that the initial conditions were very close to each other. Apparently the evolution of the system is very sensitive to initial conditions. In this case we say that the system has generated randomness. We can now see that there exist systems that, even though they can be described by simple deterministic rules, can generate randomness. Randomness generated in this way is called *chaos*. These systems are called chaotic dynamical systems, and their attractors are often called *strange* or *chaotic* attractors.

The implications of such findings are profound. If one knows exactly the initial conditions, one can follow the trajectory that corresponds to the evolution of the system from those initial conditions and basically predict the evolution forever. The problem, however, is that we cannot have perfect knowledge of initial conditions. Our instruments can only measure approximately the various parameters (temperature, pressure, etc.) that will be used as initial conditions. There will always be some deviation of the measured from the actual initial conditions. They may be very close to each other, but they will not be the same. In such a case, even if we completely know the physical laws that govern our system, due to the nature of the underlying attractor the actual state of the system at a later time can be totally different from the one predicted. Due to the nature of the system,

a   b

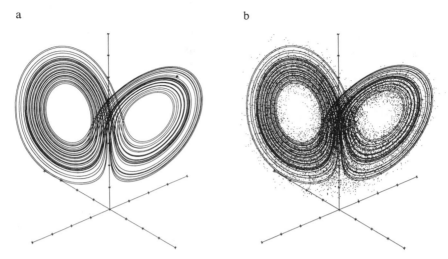

FIGURE 31. (a) An example of a strange attractor with implications in the weather fore-casting problem. This structure in the state-space represents the attractor of a fluid flow that travels over a heated surface. All trajectories (which will represent the evolution of that system for different initial conditions) will eventually converge and remain on that structure. However, any two initially nearby trajectories in the attractor do not remain nearby but diverge. (b) The effect of the divergence of initially nearby trajectories in the attractor: The dot in (a) represents 10,000 measurements (initial conditions) that are so very close to each other that they are practically indistinguishable. If we allow each of these states to evolve according to the rules, because their trajectories diverge irregularly, after a while their states can be practically anywhere. (Figure courtesy of Dr. James Crutchfield.)

initial errors are amplified and, therefore, prediction is limited. By the way, the spectra shown in Fig. 28 correspond to a time series of the variable x of the Lorenz system.

2. SOME OTHER EXAMPLES OF DISSIPATIVE SYSTEMS EXHIBITING STRANGE ATTRACTORS

1. One of the most celebrated simple dynamical systems that exhibit a strange attractor is the Hénon mapping.[101] It is described by the following equations:

$$x_{t+1} = 1 - \alpha x_t^2 + y_t$$
$$y_{t+1} = bx_t \tag{5.2}$$

For $a = 1.4$ and $b = 0.3$ the system is chaotic. The corresponding attractor is shown in Fig. 32a. The remaining figures are magnifications of the regions indicated by the square. These enlargements illustrate the fractality of the strange attractor. The same fine structure is revealed at all scales.

2. Another famous chaotic dynamical system is the Rössler system, which is described by the following equations:[176]

$$x = -y - z$$

$$y = x + ay$$

$$z = bx - cz + xz \qquad (5.3)$$

For $a = 0.38$, $b = 0.3$, and $c = 4.5$ the system is chaotic, exhibiting the attractor in Fig. 33. As we saw in Chapter 2, for these choices of constants

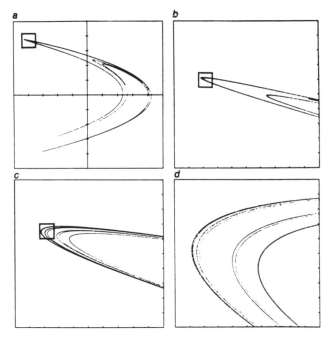

FIGURE 32.   (a) The attractor of the Hénon map with $a = 1.4$ and $b = 0.3$. (b)–(d) magnification of regions enclosed by the square in the preceding figure. Note the "fractality" (similar structure at all scales) of the attractor. (Reproduced with permission from Crutchfield et al.[37].)

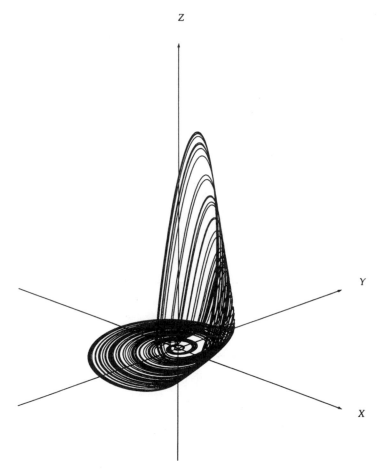

FIGURE 33.   The Rössler attractor with $a = 0.38$, $b = 0.3$, $c = 4.5$ (Figure courtesy of Dr. G. Nicolis.)

the fixed point $\bar{x} = 0$, $\bar{y} = 0$, $\bar{z} = 0$ is unstable. The trajectories near the fixed point are repelled from it along a two-dimensional surface of the phase space in which the fixed point behaves like an unstable focus.

3. The damped, periodically forced nonlinear oscillator with displacement x is described by the equation

$$\ddot{x} + k\dot{x} + x^3 = B \cos t \tag{5.4}$$

In this form the system is nonautonomous, but we can make it autonomous by writing Eq. (5.4) as

$$\dot{x}_1 = x_2$$

$$\dot{x}_2 = -kx_2 - x_1^3 + B\cos t$$

$$\dot{t} = 1$$

Thus, the phase space is three dimensional (x, y, t) or $(x, y, \cos t)$. The system exhibits very interesting dynamics ranging from periodic to chaotic. For $k = 7.5$ and $B = 0.05$ the system is chaotic. The corresponding attractor is shown in Fig. 34.

4. The following system[168]

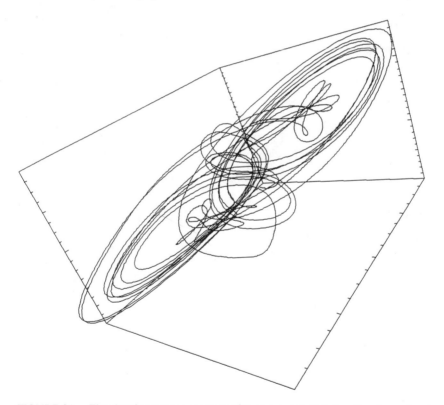

FIGURE 34. The chaotic attractor corresponding to a system that describes damped periodically forced nonlinear oscillations.

$$x = \sin ay - z \cos bx$$
$$y = z \sin cx - \cos dy$$
$$z = e \sin x \tag{5.5}$$

for $a = 2.24$, $b = 0.43$, $c = -0.65$, $d = -2.43$, and $e = 1.0$ exhibits a very

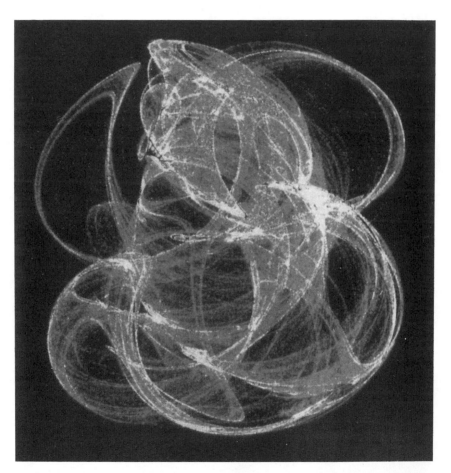

FIGURE 35.   An interesting chaotic attractor of the system described by the set of Eqs. (5.5). Green indicates regions in the attractor that are visited more often. (Reproduced with permission from Dr. C. Pickover, IBM. This figure appears in his book *Computers Pattern, Chaos, and Beauty*, St. Martin Press, 1990.) For a color reproduction of this figure see the color plates beginning facing page 148.

beautiful chaotic attractor shown in Fig. 35. This figure helps to illustrate that inside the strange attractor there exist favored regions that are visited more frequently than others (those regions are colored green).

3. DELINEATING AND QUANTIFYING THE DYNAMICS

3.1. Poincaré Sections

The most straightforward way to delineate some of the underlying dynamics is by producing the attractor.

Another very convenient way to delineate the dynamics of a system is given by the Poincaré sections or maps. A Poincaré section is a "slice" through the attractor. For an m-dimensional attractor this slice can be obtained from the intersections of a continuous trajectory with an $(m - 1)$-dimensional surface in the phase space. Thus, the system is "checked" every full orbit around the attractor. If we are dealing with a periodic evolution of period n, then this sequence consists of n dots repeating indefinitely in the same order (see Fig. 36). If the evolution is quasi-periodic, then the sequence of points defines a closed limit cycle (see Fig. 37). If the evolution is chaotic, then the Poincaré section is a collection of points that show interesting patterns with no obvious repetition (see Fig. 38). The process of obtaining a Poincaré section corresponds to sampling the state of the system occasionally instead of continuously. In many cases the appropriate sampling interval can be defined so that it corresponds to a physically meaningful measure of the dynamical system. For example, for a periodically forced oscillator, we may "mark" the trajectory at times that are multiple integers of the forcing period. Then a sequence of strictly comparable points is accumulated.

Again, here, a periodic evolution of period n manifests itself as a sequence of n points, and a quasi-periodic motion results in a continuous closed cycle. A chaotic evolution results in some interesting pattern, often revealing the fractal nature of the underlying attractor (Fig. 39).

Poincaré sections reduce the dynamics by one dimension (an n-periodic trajectory of dimension 1 becomes a set of n points that have dimension 0, and a two-dimensional torus reduces to a one-dimensional closed core). Poincaré sections thus make life a little easier while providing all the information we need when we are interested in the dynamics of the system only at specific moments in its evolution.

When dealing with attractors embedded in a 3D phase space and the

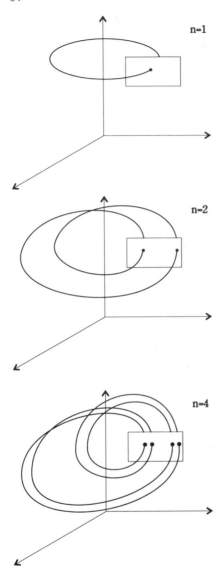

FIGURE 36. If an evolution is of period 1, then a Poincaré section is one point, if the evolution is of period 2 then the Poincaré section is two points, and so on.

Poincaré map (or section) is approximately one dimensional, we can approximately determine one coordinate if we know the other. For example, in Fig. 40, the section of the Rössler attractor with the plane $y = 0$ is shown. It looks very simple (like a one-dimensional curve). Approximating the

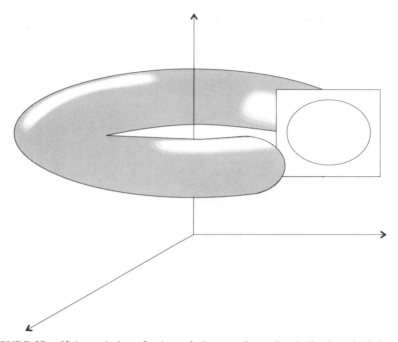

FIGURE 37. If the evolution of a dynamical system is quasi-periodic, then the Poincaré section is a sequence of points defining a closed limit cycle.

section with a simple equation allow us to determine z if we supply x. Situations like this also allow us to approximate the two-dimensional section by a one-dimensional mapping by simply recording for successive intersections the corresponding values of, say, x and plotting x_t versus x_{t+1}. We thus obtain the *return map*. The return map obtained from Fig. 40 is shown in Fig. 41.

3.2. Dimensions and Lyapunov Exponents

The dynamics of a system are dictated by the geometry of the phase space and its attractor. This geometry can be quantified by a series of dimensions and Lyapunov exponents. We have defined in Chapter 4 the topological, Euclidean, and fractal dimensions. Next we continue by presenting some other dimension definitions.

Consider a set of M points embedded in an n-dimensional Euclidean space. Also consider that this set is inside some uniform grid of size l on

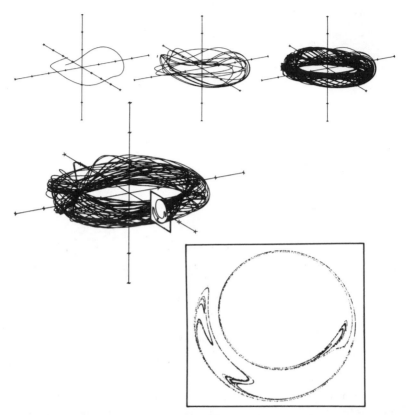

FIGURE 38.   If the evolution of a dynamical system is chaotic, the Poincaré section is a
collection of points that show interesting patterns with no apparent repetition. (Reproduced
by permission from Dr. James Crutchfield. This figure appears in the book *Chaos* by James
Gleick, Viking, USA, 1987.)

a side. Some of the n-dimensional cubes of size l may include points, and
some may not. We can define the probability of finding a point in the ith
n-dimensional cube to be $P_{li} = N_{li}/M$, where N_{li} is the number of points
in the ith cube. The average probability for a given covering, $\langle P_{li} \rangle$, is

$$\langle P_{li} \rangle \underset{l \to 0}{\approx} l^D \qquad (5.6)$$

Equation (5.6) relates the first moment (the mean) of the variable P_{li}.

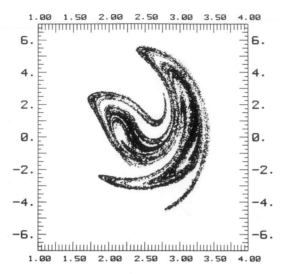

FIGURE 39. A Poincaré section of the periodically forced system exhibiting the attractor shown in Fig. 34. The section is obtained by marking the trajectory at times that are multiple integers of the forcing period.

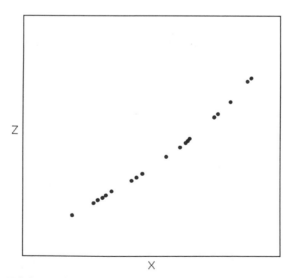

FIGURE 40. This is a section of the Rössler attractor with the plane $y = 0$. Since it looks like a 1D curve, we can approximate it by a simple equation.

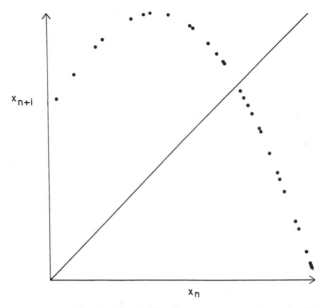

FIGURE 41.   Because of the situation in Fig. 40, we may approximate the 2D section by a 1D mapping by simply recording for successive intersections the corresponding values of, say, x and by plotting x_t vs. x_{t+1}. This way the return map (shown here) is obtained.

We can extend the arguments behind the formulation of Eq. (5.6) to consider any moment of the measure P:

$$\langle P_{li}^{(q-1)} \rangle \underset{l \to 0}{\approx} l^{(q-1)D_q} \tag{5.7}$$

From Eq. (5.7) it follows that

$$D_q \approx \lim_{l \to 0} \frac{1}{q-1} \frac{\log\langle P_{li}^{(q-1)} \rangle}{\log l}$$

For $q = 1$ a value for D_1 can be defined by taking the limit of the right-hand side as $q \to 1$. Thus,

$$D_1 \approx \lim_{q \to 1, l \to 0} \left[\frac{1}{q-1} \frac{\log \langle P_{li}^{(q-1)} \rangle}{\log l} \right]$$

$$\approx \lim_{l \to 0} \langle P_{li} \log P_{li} \rangle$$

$$\approx \lim_{l \to 0} \sum_i P_{li} \log P_{li}$$

The exponent D_1 is called the *information dimension*, because $-\sum P_{li} \log P_{li}$ is equal to the entropy $H(l)$ associated with the given distribution of probabilities and, in information theory, the rate at which new information is acquired is

$$\lim_{l \to 0} \left(\frac{-H(l)}{\log l} \right)$$

For $q = 0$,

$$D_0 \approx \lim_{l \to 0} - \frac{\log \langle P_{li}^{-1} \rangle}{\log l} \tag{5.8}$$

In Eq. (5.8), $\langle P_{li} \rangle$ is the average probability given by $[1/N(l)] \sum_i P_{li}$, where $N(l)$ is the number of hypercubes in the covering that are not empty. Thus, $\langle P_{li} \rangle = 1/N(l)$. Therefore, Eq. (5.8) becomes

$$D_0 \approx \lim_{l \to 0} - \frac{\log N(l)}{\log l}$$

which is the fractal dimension as defined by box counting (or the capacity dimension).

The capacity or fractal dimension is usually identified with the Hausdorff-Besicovitch dimension. Nevertheless, there is a fine distinction between them. The Hausdorff-Besicovitch dimension is obtained from covering the set minimally with hypercubes that may be different in size. The capacity or fractal dimension involves the same process except that the size of the hypercubes is the same. [50]

For $q = 2$ we have

$$\langle P_{li} \rangle \underset{l \to 0}{\approx} l^{D_2} \tag{5.9}$$

which is Eq. (5.6) with $D = D_2$. The exponent D_2 is called the *correlation dimension*. Consider the function

$$C(r) = \lim_{r \to 0, M \to \infty} \frac{1}{M^2} \sum_{i,j} H(r - \|x_i - x_j\|)$$

where $H(x)$ is the Heaviside step function. The summation counts the number of pairs of points in the attractor (x_i, x_j) for which the distance $\|x_i - x_j\|$ is less than r. The function $C(r)$ is called the correlation function. We can approximate $C(r)$ with $C(l)$:

$$C(l) = \lim_{M \to \infty, l \to 0} \frac{1}{M^2} \sum_{i=1}^{N(l)} M_{li}^2$$

where we have replaced the number of pairs with distance less than l by the number of pairs that fall into the same cell i of length l that is equal to M_{li}.[87] Thus,

$$\frac{1}{M^2} \sum_{i=1}^{N(l)} M_{li}^2 = \sum_{i=1}^{N(l)} P_{li} = N(l)\langle P_{li}^2 \rangle = \frac{1}{N(l)} = \langle P_{li} \rangle$$

where we have assumed that for uniform coverage $P_{li} = 1/N(l)$. Consequently, $C(l) = \langle P_{li} \rangle$; therefore, using Eq. (5.9) we get

$$C(l) \underset{l \to 0}{=} l^{D_2}$$

or

$$D_2 \underset{l \to 0}{=} \frac{\log C(l)}{\log l}$$

Extending this procedure to higher moments, we obtain an infinite number of generalized dimensions, $D_3, D_4, \ldots, D_n$. In general, $D_0 > D_1 > D_2 \cdots > D_n$, where the inequality is replaced by the equality only in special cases.[102] Like the various moments used in statistics to characterize the distribution of a random variable, the generalized dimensions can be used to give a statistical characterization of the multiple scaling in fractals. For simple fractals like the Koch curve there is just one scaling. Thus, all the generalized dimensions are the same. In more typical fractals, however, there is multiple scaling, which stems from the fact that most dynamical systems have different derivatives at different points in the flow. Thus, the folding (which, as shown in the last section, is responsible for the fractal structure of the attractor) is different from place to place. In this case the fractal is called nonuniform, and the corresponding generalized dimensions are usually different. In general, simple self-similarity or single scaling is the exception rather than the rule.

Another set of exponents that can characterize the properties of an attractor of a dynamical system is the Lyapunov exponents. The Lyapunov exponents are related to the average rates of convergence and/or divergence of nearby trajectories in phase space, and, therefore, they measure how predictable or unpredictable the system is. To introduce the idea behind Lyapunov exponents, we have produced Fig. 42. In this figure we assume that the phase space is three dimensional (however, generalization can be made for higher-dimensional phase spaces.) Let us start on the left column with a 3D volume of initial conditions (A). If this volume as it evolves contracts in only one direction, we end up with a 2D plane (B). Therefore, we say that the rate of divergence of nearby trajectors is zero along two directions and negative along the third (since the volume contracts). Therefore the spectrum of the Lyapunov exponents in this case is (0, 0, −). If we now move across the figure, we may construct from that plane something like a surface of a doughnut (torus). Since we do not disconnect what was connected (such as making a cut or a hole) nor connect what was not (such as joining the ends of the previously unjoined string or filling in the hole), we say that the plane and the torus are topologically equivalent. Thus, systems that have a torus as their attractor exhibit the same Lyapunov exponent spectrum (0, 0, −). If we now squeeze the 2D plane along one direction, we end up with a 1D line (C). Thus, when we go from a 3D cube to a 1D line, the rate of divergence of nearby trajectories is zero along only one direction while it is negative along the other two directions. From a straight-line segment we can easily produce a circle. The circle and the straight line are topologically equivalent, so systems that exhibit a limit cycle as their attractor exhibit a (0, −, −) Lyapunov exponent spectrum. If we now proceed by squeezing the two ends of the straight-line segment, we end up with a point. During a process where the initial cube of initial conditions contracts to a point, we have negative divergence along all di-

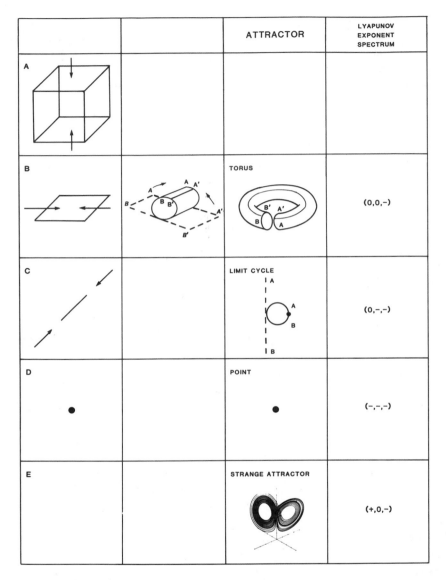

		ATTRACTOR	LYAPUNOV EXPONENT SPECTRUM
A			
B		TORUS	(0,0,−)
C		LIMIT CYCLE	(0,−,−)
D		POINT	(−,−,−)
E		STRANGE ATTRACTOR	(+,0,−)

FIGURE 42.   An introduction to Lyapunov exponents (see text).

rections, so systems that possess point attractors exhibit a Lyapunov exponent spectrum $(-, -, -)$. For chaotic dynamics the underlying attractor is not a topological manifold. In this case the Lyapunov exponent spectrum is $(+, 0, -)$. In mathematical language we can summarize the above as follows: Consider a cube of initial conditions in phase space. This cube defines a compact space M. If the dynamical system in question has an attractor that is a point Q, then $\lim_{t \to \infty} f^t X = Q$, where f^t stands for the time evolution of any system and X is a set of initial conditions. Thus, the initial three-dimensional volume contracts to a zero-dimensional point. In order for that to happen the volume must contract along all three directions. In this case there exist three Lyapunov exponents that are all negative, since the volume (or nearby trajectories) contracts in all directions.

If the system in question possesses a limit cycle as its attractor, then $\lim_{t \to \infty} f^t X = \Gamma$. Since Γ is a closed trajectory having a topological dimension of 1, the initial volume must contract along two directions. It does not contract along the direction of the orbit. Thus, in this case two exponents are negative and one is zero.

If the system has a torus as an attractor, then $\lim_{t \to \infty} f^T X = \Delta$. Accordingly, since Δ is a surface having a topological dimension of 2, the initial volume must contract along one direction only. Each orbit, corresponding to a fixed initial condition, will fill the torus as $t \to \infty$. Orbits that correspond to two different initial conditions may reach that surface at different points, but once in the attractor their separation remains constant (the flow f^t is the same for all initial conditions). This means that nearby trajectories do not diverge or converge. Thus, in this case two exponents are zero, and one is negative.

If the system possesses a nontopological chaotic attractor S, then $\lim_{t \to \infty} f^t X = S$. In this case the initial volume contracts along some direction to count for the approaching of a three-dimensional volume to a lower-dimension submanifold. Along the direction of the trajectory it will not contract. However, the chaotic attractor has the property that the nearby trajectories do not stay nearby as $t \to \infty$, but they diverge. One may imagine a compact volume M' of measurements in the attractor, which under this action will expand. Thus, in this case one exponent must be positive, one must be zero, and the other one must be negative.

Following the above, we now give a formal definition of Lyapunov exponents. We start with a set of initial conditions in an attractor (embedded in n-dimensional Euclidean space) that are confined within an n-dimensional sphere. Subsequently, we begin to monitor the long time evolution of this sphere. We order the principal axes of this sphere from the most rapidly to the least rapidly growing, and we compute the mean growth rate λ_i of any given principal axis p_i. We may define these growth rates as follows:

$$\lambda_i = \lim_{T \to \infty} \frac{1}{T} \int_0^T dt \, \frac{d}{dt} \ln \left[\frac{P_i(t)}{P_i(0)} \right]$$

$$= \lim_{T \to \infty} \frac{1}{T} \ln \left[\frac{P_i(T)}{P_i(0)} \right]$$

Here $p_i(0)$ is the radius of the principal axis p_i at $t = 0$ (i.e., in the initial hypersphere), and $p_i(T)$ is its radius after a long time T. The set of λ_i's is referred to as the Lyapunov exponent spectrum. Apparently there are as many Lyapunov exponents as the dimension of the phase space.

When at least one Lyapunov exponent is positive, then the system at hand is chaotic, and the initial sphere will evolve to some complex ellipsoid structure reflecting the exponential divergence of nearby initial conditions along at least one direction on the attractor. This "sensitivity" to the initial conditions results in an inability to predict the evolution of the trajectory beyond an interval of time approximately the inverse of the divergence rate. When no positive Lyapunov exponent exists, then no exponential divergence exists, and thus the long-term predictability of the system at hand is guaranteed.

In one sense, we may say that the exponents measure the rate at which the system destroys information. Positive. exponents give an idea of how fast information contained in a set of initial conditions, initially very close to each other, is lost due to the action of the stretching and folding of the chaotic attractor. Negative exponents give an idea about the average rate at which information contained in transients is lost. Thus, in a sense, negative and positive exponents define the dominant time scales in the evolution of a dynamical system (see, for example, Keppenne and Nicholis[117]).

4. DETERMINING THE VARIOUS DIMENSIONS AND LYAPUNOV EXPONENTS FOR A SYSTEM OF ORDINARY DIFFERENTIAL EQUATIONS

Knowing the mathematical formulation of a dynamical system allows us to generate the underlying attractor by numerical integration of the governing equations. That produces a set of points in the phase space. Then a straightforward application of the box-counting approach or other relevant approaches discussed previously yields the fractal dimension, the correlation dimension, and so forth.

Following the definition of the Lyapunov exponents, to estimate the exponents one simply has to consider nearby points and monitor them as they are allowed to move in the attractor according to the governing equations. Based on this idea, Benettin *et al.*[17] developed a simple procedure that measures expansion by introducing small perturbations about a fiducial trajectory found through integration of the equations of motion. At first a point $x(0)$ on the attractor is selected, and then a nearby point $x'(0)$ is located (see Fig. 43). The two points define an initial vector d_{i1}. Then the equations are solved, and points $x(0)$ and $x'(0)$ are carried to positions $x(1)$ and $x'(1)$. Let us denote the time elapsed as T, and the vector between those two points as d_{f1} (f stands for final). If the system is chaotic, then point $x'(0)$ evolves in the direction of the greatest expansion at $x'(1)$. The elapsed time T is kept short to guarantee that the separation does not become comparable to the size of the attractor. At time T a point $x''(1)$ near to $x(1)$ is located, which lies along the line connecting $x(1)$ and $x'(1)$, and the process is repeated. Successive iterations pull the line segment along the direction of greatest instability. The largest Lyapunov exponent is then estimated from the average difference in the initial and final vector magnitudes:

$$\lambda = \lim_{n \to \infty} \frac{1}{nT} \sum_{j=1}^{n} \ln\frac{d_{fj}}{d_{ij}}$$

where n is the number of iterations.

This approach, however, provides only the largest Lyapunov exponent. A method for estimating the entire Lyapunov spectrum has its roots in the discussion in Chapters 2 and 3. From Chapter 2 we know that the linearized equations describe the *local dynamics* via the evolution of small perturbations

$$\dot{\mathbf{x}}' = A\mathbf{x}' \qquad \text{or} \qquad \mathbf{x}'(t) = e^{tA}\mathbf{x}'(0)$$

where $\mathbf{x}'(0)$ is the initial vector. Recall from Chapter 3 [Eq. (3.14) and subsequent discussion] that the matrix A can be used to find the entire Lyapunov spectrum. In fact, the Lyapunov spectrum is the set of the eigenvalues of A that can be estimated at any point along the numerically integrated trajectory if the governing equations are known.

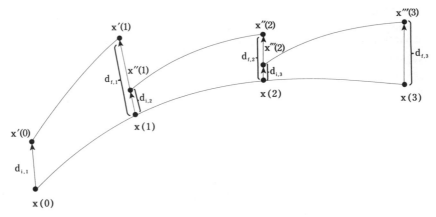

FIGURE 43.   The procedure behind calculating Lyapunov exponents from a system of differential equations.

The Lyapunov spectrum is closely related to the dimension of the associated attractors. It has been conjectured [69,124] that the information dimension D_1 is related to the Lyapunov spectrum via the equation

$$D_1 = j + \sum_{i=1}^{j} \lambda_i \bigg/ |\lambda_{j+1}| \qquad (5.10)$$

where j is defined by the condition that

$$\sum_{i=1}^{j} \lambda_i > 0 \qquad \text{and} \qquad \sum_{i=1}^{j+1} \lambda_i < 0$$

For example, for the Lorenz system $\lambda_1 = 2.16$, $\lambda_2 = 0.00$, and $\lambda_3 = -32.4$. Thus, $\lambda_1 + \lambda_2 = 2.16 > 0$ and $\lambda_1 + \lambda_2 + \lambda_3 = -30.24 < 0$. Therefore, $j = 2$. According to Eq. (5.10), $D_1 = 2 + 2.16/32.4 \approx 2.07$, which is the value obtained from box counting.

As mentioned, the flow within the attractor may not be the same everywhere. As a result, the rate of expansion (or contraction) may not be constant. There may be regions in the attractor where the expansion (or contraction) is greater than in other regions. Thus, to estimate the average rates, we must perform many iterations. Figure 44 shows the positive Lyapunov exponent for the familiar Lorenz attractor. Certain regions exhibit greater expansion than other regions. Recall that the predictability of the

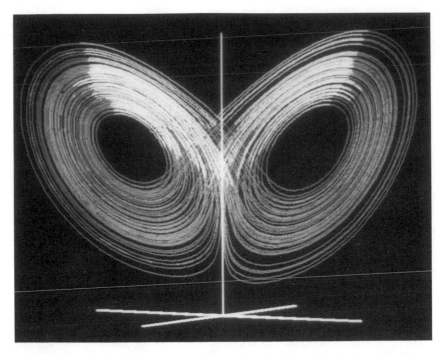

FIGURE 44.   The largest Lyapunov exponent value along a path in the Lorenz attractor. Green represents largest local divergence rate less than zero, and red represents largest local divergence rate greater than zero. (Figure courtesy of Dr. J. Nese.) For a color reproduction of this figure see the color plates beginning facing page 148.

system depends on the rate at which nearby trajectories diverge. Therefore, regions that correspond to smaller expansion rates could be identified as regions of greater predictability within the chaotic flow!

5. MULTIPLE ATTRACTORS AND FRACTAL BASINS

A dynamical system may not have just one attractor for a given set of parameter values. It may have two, three, or more. Which attractor will be chosen depends simply on the initial condition. In such cases the phase space is divided into basins of attraction. Figures 45 shows basins of attraction of a simple hypothetical system that exhibits two attractors A and B. For illustration purposes the phase space is divided into two basins by a boundary C. All evolutions initiated from an initial condition to the

STATE SPACE

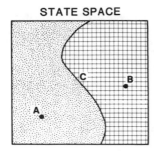

FIGURE 45.   Illustration of basins of attraction and basin boundary. In this hypothetical situation a dynamical system exhibits two attractors A and B. The phase space is divided into basins by boundary C. Evolutions from an initial condition to the right of C converge to attractor B, while evolutions from an initial condition to the left of C converge to attractor A. (Tsonis and Elsner.[212])

right of C will converge and stay on attractor B, and all evolutions initiated from an initial condition to the left of C will converge and stay on attractor A. In general, the "shape" of C is not that simple, and often C is a fractal set. To illustrate this possibility, consider the dynamical system

$$\ddot{x} + k\dot{x} + \Omega \sin x = f \cos \omega t. \tag{5.11}$$

Equation (5.11) describes the motion of an externally forced ($f \cos \omega t$), damped ($k\dot{x}$), nonlinear ($\Omega \sin x$) pendulum. The parameters f, ω, k, and Ω refer to the amplitude of the external forcing, the frequency of the external forcing, the coefficient of the friction, and the frequency of the natural oscillation, respectively.

For $k = 0.2$, $f = 2.0$, $\omega = 0.997331$, and $\Omega = 1.0$, the system possesses two periodic attractors, say A and B. The basins of attraction (for part of the available phase space) are shown in Fig. 46. These basins are generated as follows: A grid of 400 × 400 initial conditions [$y(1) = x$, $y(2) = \dot{x}$] is

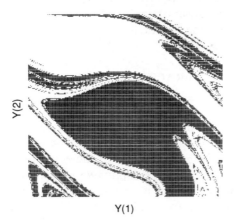

Y(2)

Y(1)

FIGURE 46.   Basins of attraction of the system described by the equation $\ddot{x} + k\dot{x} + \Omega \sin x = f \cos \omega t$. For $k = 0.2$, $f = 2.0$, $\omega = 0.997331$, and $\Omega = 1.0$, the system exhibits two periodic attractors, A and B. Every initial condition converging on attractor A is indicated by a black dot. Thus, black-and-white regions are pictures of the basins of attraction. Note the complex fractal field with fine-scale structure.

considered, and Eq. (5.11) is integrated for each initial condition until it is close to one of the two attractors. If the orbit goes to attractor A, a black dot is plotted at the corresponding initial condition. If the orbit goes to attractor B, nothing is plotted. Thus, black-and-white regions are pictures of the basins of attraction to the accuracy of the grid of the computer plotted. A rather complex fractal field with fine-scale structure is evident. An even more beautiful example of a fractal basin is shown in color in

FIGURE 47.   A colorful version of Fig. 46. It represents the basins of attraction for the damped periodically driven pendulum. This system is somewhat different in formulation than that corresponding to the basins of Fig. 46. Shades of red and blue correspond to initial conditions that approach the corresponding attractor fast (dark) and slow (light) (Figure courtesy of Dr. C. Grebogi, University of Maryland. This figure appears in *Science* **238**, 1987, copyright 1987 by the AAAs.) For a color reproduction of this figure see the color plates beginning facing page 148.

Fig. 47. Fractal basin boundaries have been observed in many dynamical systems.[89,90,96,149]

As pointed out by Grebogi et al.,[89] a fractal basin boundary causes high sensitivity to initial conditions even if all solutions are found to be periodic. This sensitivity is not a result of the action of a chaotic attractor, but is a result of the fine-scale structure of the basin, which makes very close initial conditions converge on two different attractors. Thus, even if the two attractors are periodic (i.e., they guarantee long-term predictability), the system may be unpredictable if an initial condition is not known exactly. In effect, the uncertainty on the initial condition plates the system on a position of the basin from which it is attracted to the opposite attractor!

To this end we consider an initial condition $(x, \dot{x})$ and integrate Eq. (5.11) for initial conditions $(x, \dot{x})$, $(x, \dot{x} + \epsilon)$, and $(x, \dot{x} - \epsilon)$, where ϵ is a small perturbation, until they approach one of the attractors. If either or both of the perturbed initial conditions do not approach the same attractor that the unperturbed initial condition approaches, then we consider $(x, \dot{x})$ as uncertain. For numerous initial conditions the procedure yields the fraction f of uncertain initial conditions. This fraction scales as $f \sim \epsilon^a$ where a is related to the fractal dimension D of the basin via the equation

$$a = d - D$$

where d is the Euclidean dimension of the phase space. For the system described by Eq. (5.11) we find that $a \sim 0.30$ and, thus, $D \sim 1.70$. Since $f^{1/a} \sim \epsilon$, it follows that, if we want to improve our uncertainty in predicting the correct attractor by a factor of 2, then we must improve our accuracy of measurement by a factor of $2^{1/a} = 2^{1/0.3} \sim 10$. Hence, fractal basin boundaries represent an obstruction to predictability.

6. AN OBVIOUS QUESTION

When a system exhibits a strange or chaotic attractor, the trajectory that describes its evolution is of infinite length, never crosses itself, and occupies zero volume in the phase space. At the same time, nearby trajectories diverge. This exponential divergence is a local feature. This is so because attractors have finite size, and thus two orbits on a chaotic attractor cannot diverge forever. How can we then get local expansion and global boundedness at the same time? As illustrated in Fig. 48, a volume element under the action of a chaotic flow will at first be pulled along the direction

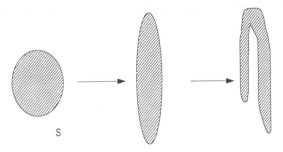

FIGURE 48. Evolution of a volume element *S* under the action of a chaotic flow. The element is stretched along the direction of greatest instability, but boundedness of dynamics prevents unlimited expansion by folding. Successive iterations result in the asymptotic attractor possessing a fractal structure.

of the greatest instability. This *stretching,* however, cannot occupy more and more space since the attractor is of a finite size. The mechanism that prevents that is *folding.* Eventually the attractor must fold over onto itself. Successive iterations of this process result in the asymptotic attractor with folds within folds *ad infinitum* (i.e., a fractal object).

Although two orbits diverge locally everywhere within the attractor, they eventually pass close to one another at some time in the future. We might think of the action of a chaotic attractor as a mixing procedure according to which orbits are mixed in a similar way that a baker mixes bread dough by kneading it. If a drop of black ink is dropped in the dough, the kneading, a combination of rolling the dough and folding it over, will eventually mix the ink completely within the dough. Chaos works in the same way, mixing the state-space.

Up to now we have discussed several dissipative dynamical systems that exhibit chaos. Those systems, however, do not exhibit chaos for all choices of their parameters. For example, the Lorenz system has two periodic attractors for $a = 10$, $b = 24.06$, and $c = \frac{8}{3}$. Next we address the mechanisms via which a dissipative system changes its character as some or all of its parameters are varied.

CHAPTER 6

BIFURCATIONS AND ROUTES TO CHAOS

1. HOPF BIFURCATION

In Chapter 5 we saw that the stability of a dynamical system depends on the values that certain parameter(s) assume. Thus, the equilibrium state of the system may be stable for some range of the parameter(s) and unstable for others. When the equilibrium state is stable, we need not worry about anything. No changes are going to take place in the dynamics of the system. If the equilibrium state is unstable, however, changes are anticipated. We saw that in these cases the equilibrium state is a repeller or a saddle, and all trajectories diverge. Where do they go? What happens to them?

We begin to address this issue with a simple dynamical system described by the equations[15]

$$\dot{x}_1 = x_2 + x_1\mu(1 - x_1^2 - x_2^2)$$
$$\dot{x}_2 = -x_1 + x_2\mu(1 - x_1^2 - x_2^2) \qquad (6.1)$$

The equilibrium state of the system is $\bar{x}_1 = 0, \bar{x}_2 = 0$. We may now linearize this system by taking $x_1 = \bar{x}_1 + x_1'$ and $x_2 = \bar{x}_2 + x_2'$, where x_1' and x_2' are very small fluctuations about the equilibrium state.

After linearization we have

$$\dot{x}_1' = \mu x_1' + x_2'$$
$$\dot{x}_2' = -x_1' + \mu x_2'$$

or, in vector notation,

$$\dot{\mathbf{x}}' = A\mathbf{x}'$$

where

$$A = \begin{pmatrix} \mu & 1 \\ -1 & \mu \end{pmatrix}$$

The eigenvalues λ_1 and λ_2 are found from $\mathrm{Det}(A - \lambda I) = 0$ to equal $\mu + i$ and $\mu - i$, respectively. Thus, for $\mu < 0$ the equilibrium state is locally stable, and for $\mu > 0$ it is locally unstable. For $\mu = 0$ the eigenvalues are imaginary, so the equilibrium state is a center and no judgment about the stability can be made. The interesting observation, however, is that as μ changes so does the stability.

We now address the important question of what really happens as the stability of the system changes. If we let $x_1 = (r/\mu)\cos\theta$ and $x_2 = (r/\mu)\sin\theta$, we can obtain, after some manipulation of Eq. (6.1), the equations

$$x_1\dot{x}_1 + x_2\dot{x}_2 = \dot{r}r = \mu r^2(1 - r^2)$$

$$x_2\dot{x}_1 - x_1\dot{x}_2 = -r^2\dot{\theta} = r^2 \tag{6.2}$$

or

$$\dot{r} = \mu r(1 - r^2)$$

$$\dot{\theta} = -1 \tag{6.3}$$

The first equation of system (6.3) can be solved analytically. By letting $y = 1/r$, we obtain

$$\dot{y} = \frac{\mu(1 - y^2)}{y}$$

or

$$\frac{y\,dy}{1 - y^2} = \mu\,dt \tag{6.4}$$

Integrating Eq. (6.4) from an initial condition $y_0 = y(0)$ to some later state $y(t)$, we obtain

$$\ln\frac{1 - y_0^2}{1 - y^2(t)} = 2\mu t$$

or, by substituting $y = 1/r$,

$$r^2(t) = \frac{r_0^2}{r_0^2 + (1 - r_0^2)e^{-2\mu t}} \tag{6.5}$$

Equation (6.5) gives the evolution of the system from some initial state as a function of time. If $\mu < 0$, then, as $t \to \infty$, $r(t) \to 0$. Since $r^2 = x_1^2 + x_2^2$, $r(t) \to 0$ means that $x_1 \to 0$ and $x_2 \to 0$; i.e., the system evolves toward the stable equilibrium state. If $\mu > 0$, then, as $t \to \infty$, $r(t) \to 1$. Thus, as $t \to \infty$, $x_1^2(t) + x_2^2(t) \to 1$. In a coordinate system (x_1, x_2), this is the equation of the unit circle. Therefore, the system evolves to a state described in the $x_1 x_2$ plane by a circle that is a *limit* circle (Fig. 49). Obviously, the final state is now periodic. This limit cycle is a stable closed orbit. (The significance of the second equation ($\dot{\theta} = -1$) is that the trajectory describing the evolution of the system spirals clockwise toward the unit limit circle). Therefore, we see that as soon as μ becomes greater than zero a significant change in the dynamics of the system is observed. Local instability gives way to a globally stable closed orbit. Such a change in the

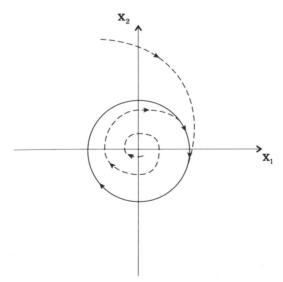

FIGURE 49.   The unit cycle as a limit cycle of Eq. (6.1).

dynamical behavior of the system is called a *bifurcation*. In our example the critical value of zero of the parameter μ is called a *Hopf bifurcation point*, and the bifurcation is called *Hopf bifurcation*.[104,143]

Before we continue with our next example, we mention the following points: (1) Since the origin is unstable (once a limit cycle is born), the center of the limit cycle is a repeller. Therefore, all evolutions initiated inside the limit cycle are repelled from the origin and are attracted by the limit cycle. (2) As can be deduced from Eqs. (6.4) and (6.5), for $\mu < 0$ the point attractor is approached faster for high negative values than for small negative values. As μ approaches zero, the time that it takes to reach the point attractor tends toward infinity while the point attractor begins to "give in" to the impending limit cycle (see Fig. 50). (3) During the first excitation a point attractor "becomes" a limit cycle. We may equivalently say that we increase the (topological) dimension of the attractor by 1. In some cases it is possible to further increase the dimension of the attractor by 1, thus ending up with a torus. Such a bifurcation is called the *secondary Hopf bifurcation* or the *second excitation*. Thus, in two dimensions a limit cycle is "waiting" for the attractor of the fixed point to weaken. By analogy, in three dimensions a torus is "waiting" for the attractor of the limit cycle to weaken. This event can be generalized to the bifurcation of an n-dimensional torus into an $(n + 1)$-dimensional torus.

In a sense this example was very simple. The steady-state equilibrium observed for $\mu < 0$ gives way to a periodic equilibrium state as μ becomes positive. We find no more surprises as μ is increased further. We thus arrive at the following question. Are there systems whose dynamical behavior changes continually as its parameter(s) change? If so, what is the final fate of such systems?

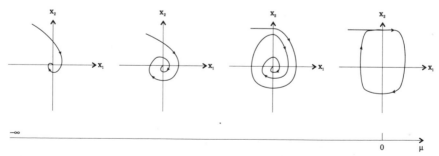

FIGURE 50.   As the controlling parameter μ is varied, the approach to the equilibrium state $(\bar{x}_1 = 0, \bar{x}_2 = 0)$ takes longer and longer. As $\mu \to 0$, this time tends to infinity, and the point attractor gives way to a periodic attractor.

1.1. When Do Limit Cycles Exist?

Consider a two-dimensional dynamical system $\dot{x} = Ax$, where the matrix A is a function of some controlling parameter μ. We suppose that for some suitable range of μ values the eigenvalues are differentiable in μ and complex: $\lambda_i(\mu) = \alpha(\mu) \pm i\beta(\mu)$, where $\alpha(\mu)$, $\beta(\mu)$ indicate the real and imaginary parts of the eigenvalues as functions of the controlling parameter μ. According to the Hopf bifurcation theorem in $\mathbf{R}^2$, a limit cycle for the dynamical system in question exists when the following conditions are satisfied:

1. $d\alpha(\mu)/d\mu > 0$
2. $\alpha(0) = 0$
3. $\beta(0) \neq 0$

This theorem, under certain conditions, can be extended to $\mathbf{R}^k$.

2. PITCHFORK BIFURCATION AND PERIOD DOUBLING

Let us now employ what is considered the simplest, but one of the most interesting, dynamical systems, the logistic map.[146]

$$x_{n+1} = \mu x_n(1 - x_n) = f(x_n) \tag{6.6}$$

This map is often used to model population dynamics. The parameter μ is the nonlinearity parameter. Figure 51 shows $f(x)$ as function of x for different values of μ. Note that $f(x)$ has one extremum (in this case a minimum) for $x = 0.5$ and that, as μ increases, the curve becomes more steeply humped (the extremum assumes higher values). Since the population has to be nonnegative, we must have $0 \leq x_n \leq 1$. Furthermore, if we require nontrivial dynamic behavior, we must have $0 < x_n < 1$, since for $x_n = 0$ or $x_n = 1$, $x_{n+1} = 0$; i.e., the population becomes extinct. Since $f(x)$ obtains a maximum value of $\mu/4$ at $x = 0.5$, the map possesses nontrivial dynamic behavior only if $\mu \leq 4$. Considering also that for $\mu < 1$ all solutions are attracted to zero (extinction), nontrivial dynamic behavior can be achieved only if $0 < x_n < 1$ and $1 \leq \mu \leq 4$.

The equilibrium (or fixed) points of the logistic equation can be found by setting $x_{n+1} = x_n = \bar{x}$. This corresponds to a situation where the pop-

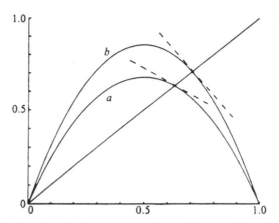

FIGURE 51.   The function $f(x_n) = x_{n+1} = \mu x_n (1 - x_n)$ for two values of the nonlinearity parameter μ. As μ increases, the curve becomes more steeply humped. Equilibrium is defined when $x_{n+1} = x_n$. Thus, in this figure the equilibrium point is the intersection of the curves with the straight line $x_{n+1} = x_n$. The dashed lines indicate the slope of the curves at the equilibrium points. (Reproduced by permission from May[146]).

ulation of the next generation, x_{n+1}, is the same as the population of the previous generation. We may say that the population repeats after one generation, or that the population repeats with period 1. In this case Eq. (6.6) can be written as

$$\bar{x} = \mu\bar{x}(- \bar{x})$$ (6.7)

The roots are 0 and $1 - 1/\mu$.

   We wish to investigate the stability of the map about the nontrivial equilibrium $\bar{x} = 1 - 1/\mu$. Again we proceed with investigating the properties of the system in the presence of very small fluctuations about $\bar{x}$. We can rewrite Eq. (6.6) as

$$\bar{x} + x'_{n+1} = \mu(\bar{x} + x'_n)(1 - (\bar{x} + x'_n))$$

or

$$\bar{x} + x'_{n+1} = \mu(\bar{x} + x'_n) - \mu(\bar{x} + x'_n)^2$$

Neglecting all terms involving fluctuations higher than first order, we obtain

$$\bar{x} + x'_{n+1} = \mu(\bar{x} + x'_n) - \mu(\bar{x}^2 + 2\bar{x}x'_n)$$

or, taking into account Eq. (6.7),

$$x'_{n+1} = \mu x'_n (1 - 2\bar{x}) \tag{6.8}$$

Recognizing that $1 - 2\bar{x} = \dot{f}(\bar{x})$ (the first derivative of $f(x)$ evaluated at $\bar{x}$), we arrive at

$$x'_{n+1} = \dot{f}(\bar{x})x'_n \tag{6.9}$$

Obviously, as $t \to \infty$, $x'_{n+1} \to 0$ as long as $|\dot{f}(\bar{x})| < 1$ or

$$-1 < \mu - 2\mu\bar{x} < 1$$

or

$$-1 < \mu - 2\mu\left(1 - \frac{1}{\mu}\right) < 1$$

or

$$-1 < \mu < 3 \tag{6.10}$$

Therefore, as long as $1 < \mu < 3$, $x_{n+1} = x_n = \bar{x}$, which means that the population remains constant in time. Figure 52 shows the evolution from two different initial conditions for $\mu = 2.707$. In this figure the system attains the dynamic behavior of no change (period 1 evolution).

Since $\dot{f}(\bar{x})$ is actually the slope of $f(x)$ at $x = \bar{x}$, we find that the condition for stability translates to having the slope at $\bar{x}$ between $\pm 45°$. This is shown in Figure 51 as a dashed line. Note that as μ increases, the slope becomes increasingly steep. For $\mu > 3$ the slope is steeper than $-45°$, and thus the fixed point becomes unstable.

What is the fate of the population now? Since the population after one generation is "departing" from the equilibrium steady state, to answer this question we must find the population after two generations. We thus seek to find the relation between x_{n+2} and x_n. According to Eq. (6.6),

$$x_{n+2} = \mu x_{n+1}(1 - x_{n+1})$$

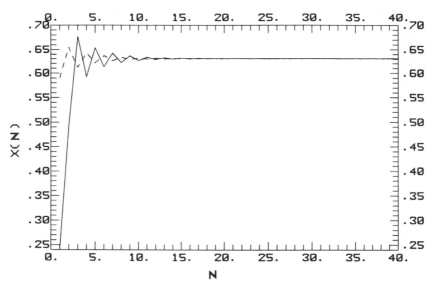

FIGURE 52.   Evolution of the logistic map with $\mu = 2.707$ from two different initial conditions. Both evolutions become identical and of period 1 (i.e., repeating every generation).

or

$$x_{n+2} = \mu^2 x_n(1 - x_n)[1 - (\mu x_n(1 - x_n))]$$

or

$$x_{n+2} = \mu f(x)(1 - f(x))$$

or

$$x_{n+2} = f[f(x)] \qquad (6.11)$$

Thus, the population after two generations is obtained by the second iterate of $f(x)$. We denote this as $f^2(x)$ and we will generalize Eq. (6.11) for the mth iterate

$$x_{n+m} = f^m(x) \qquad (6.12)$$

The graph of $x_{n+2} = f^2(x)$ as a function of x is given in Fig. 53 for different values of μ. Note that now the function has three extrema, which again

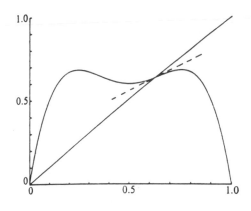

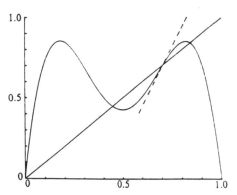

FIGURE 53.   As in Fig. 51, but now x_{n+2} vs. x_n is plotted. This function has three extremes and becomes steeper as μ increases. Note that for $\mu < 3$ (top) the line $x_{n+2} = x_n$ intersects the curve at one point, while for $\mu > 3$ (bottom) it intersects it at three points. (Reproduced by permission from May.[146])

become steeper as μ increases. Let us first explain why $f^2(x)$ should look as it appears in Fig. 53.

In this case ($m = 2$), for an initial condition x_0,

$$x_2 = f^2(x_0) \tag{6.13}$$

where

$$x_1 = f(x_0) \qquad x_2 = f(x_1)$$

Using the chain rule, we can verify that

$$\dot{f}^2 = \dot{f}(x_0)\dot{f}(x_1) \tag{6.14}$$

Thus, if $x_0 = \bar{x}$, then

$$\dot{f}^2(\bar{x}) = [\dot{f}(\bar{x})]^2 \tag{6.15}$$

We mentioned that for $x = 0.5$, $f(x)$ is extreme. Thus, $\dot{f}(0.5) = 0.0$. It then follows from Eq. (6.14) that $\dot{f}^2(0.5) = 0.0$. Therefore, f^2 at $x = 0.5$ is extreme (a minimum in this case). Are there any other extremes in f^2? By Eq. (6.14), f^2 is extreme (maximum is this case) at any x_0's that produce an x_1 equal to 0.5. In this case $\dot{f}(x_1)$ is zero, and therefore $\dot{f}^2$ is also zero. Thus, we seek initial conditions that give as a first iterate the value 0.5. There are two such initial conditions.[58] Thus, f^2 has three extremes (two maxima and one minimum).

Recalling Eq. (6.9), we have $x'_{n+1} = \dot{f}(\bar{x})x'_n$ and $x'_{n+2} = \dot{f}(\bar{x})x'_{n+1}$ or $x'_{n+2} = (\dot{f}(\bar{x}))^2 x'_n$. Thus, since $(\dot{f}(\bar{x}))^2 = \dot{f}^2(\bar{x})$, it follows that $x'_{n+2} = \dot{f}^2(\bar{x})x'_n$, which means that in this case the stability criterion is $|\dot{f}^2(\bar{x})| < 1$. Therefore, for $f^2(x)$ the stability criterion translates to requiring the angle of the slope of $x = \bar{x}$ to be between $-45°$ and $45°$. This slope is indicated in Fig. 53 by dashed lines. We also notice from Fig. 53 that the line $x_{n+2} = x_n$ intersects $f^2(x)$ at $x = \bar{x}$ for $\mu < 3$, but for $\mu > 3$ it intersects $f^2(x)$ (which by now has become steep enough) at two additional points. At the same time the slope at $x = \bar{x}$ becomes greater than $45°$. Thus, when stability gives way to instability, the stable fixed point becomes unstable and, simultaneously, two new stable fixed points $(\bar{x}_1, \bar{x}_2)$ appear. These two points are stable because the slope of $f^2(x)$ at those points is between $0°$ and $45°$. Therefore, we have three equilibrium points, two of which $(\bar{x}_1$ and $\bar{x}_2)$ *are not* fixed points of $f(x)$ [i.e., $\bar{x}_i \neq f(\bar{x}_i)$, $i = 1, 2$], but are fixed points of $f^2(x)$ [i.e., $\bar{x}_1 = f^2(\bar{x}_1)$ and $\bar{x}_2 = f^2(\bar{x}_2)$]. It thus follows that both fixed points repeat after two iterations, which in turn dictates that once $x = \bar{x}_1$ or $x = \bar{x}_2$, then the population must alternate between these two states. The population, therefore, is now periodic every two generations.

We have thus observed a *period doubling* as the parameter μ crossed the critical value $\mu = 3$. This change in dynamic behavior—the branching of one fixed point into two fixed points and one unstable, point—is called *pitchfork bifurcation* (see Fig. 54). Figure 55 shows the evolution of the system from two different initial conditions for $\mu = 3.35$. After the transients die out the system settles in a period 2 evolution.

At this point we may ask the obvious question: What happens as

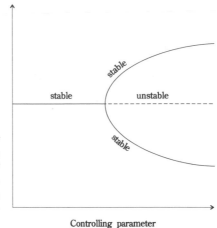

FIGURE 54. Graphical representative of pitchfork bifurcation. First we have just one stable point. Then at some critical value of the controlling parameter the stable point becomes unstable while two new stable points appear. The evolution is from periodic of period 1 to periodic of period 2.

$f^2(x)$ becomes steeper and steeper as μ keeps on increasing. Eventually the slope will fall to a point at which the two stable points become unstable. A further increase in μ results in another period doubling, as each of the two now unstable fixed points gives rise to two new stable fixed points. In

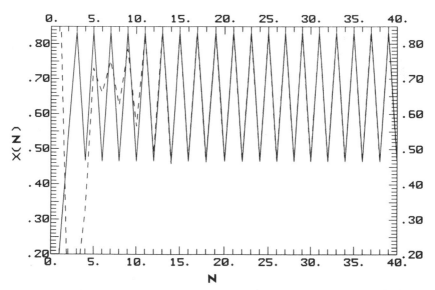

FIGURE 55. Evolution of the logistic map with $\mu = 3.35$ from two different initial conditions. Both evolutions become identical and of period 2 (i.e., repeating every two generations).

this case the evolution becomes periodic of period 4 (repeating every four generations). Figure 56 is similar to Fig. 55, but for $\mu = 3.5$. Here, after the transients die out, the system settles to a period 4 evolution. This period doubling increases indefinitely until the period becomes infinite, in which case we practically have an aperiodic deterministic evolution. This type of evolution is called *chaotic.*

Figure 57 illustrates such an evolution from two different initial conditions for $\mu = 4.0$. Note that, unlike Figs. 52, 55, and 56, the system does not settle down to some "recognizable" or predictable dynamics, as with any periodic evolution. The evolution is quite different and depends on the initial condition. Such a property is unique for chaotic systems. It is often referred to as the sensitivity to the initial conditions property. When chaotic the system exhibits an "erratic" behavior similar to those observed in random processes. As a matter of fact, the spectra of the logistic evolution for $\mu = 4$ (Fig. 58) are indistinguishable from that of a completely random sequence (Fig. 58). *The discovery of such states thus becomes extremely important in that seemingly random processes might actually be completely deterministic (chaotic).*

The period-doubling scenario can be graphically presented by the *bifurcation diagram* shown in Fig. 59. The *x* axis is the controlling parameter

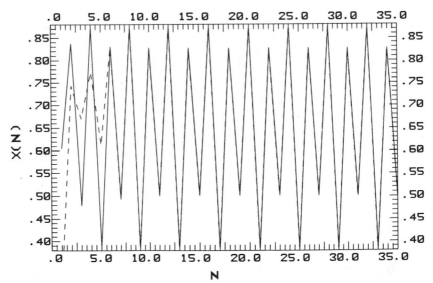

FIGURE 56. As in Fig. 55, but for $\mu = 3.5$. The evolution has now become periodic of period 4 (repeating every four generations).

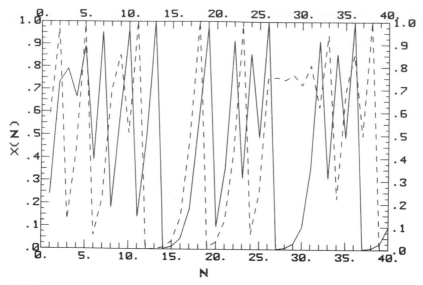

FIGURE 57.   As in Fig. 56, but for $\mu = 4.0$. In this case the evolution has become chaotic. Any two evolutions starting from two different initial conditions do not become identical after the transients die out.   Instead we have two uncorrelated evolutions.

μ. For some range $1 < \mu < \mu_1$ the equilibrium state of the system is stable, and the evolution is of period 1. At $\mu = \mu_1$ the fixed point becomes unstable, and for $\mu_1 < \mu < \mu_2$ the unstable fixed point gives way to two fixed points (a period 2 limit cycle), and so on.

3. FLIP BIFURCATION AND PERIOD DOUBLING

Once the Hopf bifurcation in a continuous system has occurred, the newly born limit cycle of period 1 is stable. In our example in Section 1 this limit cycle remains stable for all values of $\mu > 0$. There are, however, continuous systems where the period-doubling scenario (similar to that illustrated with the logistic map) will also take place.

Consider the dynamical system

$$\ddot{x} + k\dot{x} + x^3 = b \cos t \qquad (6.16)$$

This equation is called Duffing's equation, and it describes forced damped nonlinear oscillations. The (external) forcing is defined by $b \cos t$, the damping is determined by k, and the restoring nonlinear force is x^3. Physically, the damping is related to the tendency of the system to react to changes dictated by the restoring and external forcing. Such an equation can be used to model the evolution of a sinusoidally forced structure undergoing large deflections.

Equation (6.16) can be written as a first-order system:

$$\dot{x}_1 = x_2$$

$$\dot{x}_2 = -kx_2 - x_1^3 + b \cos t$$

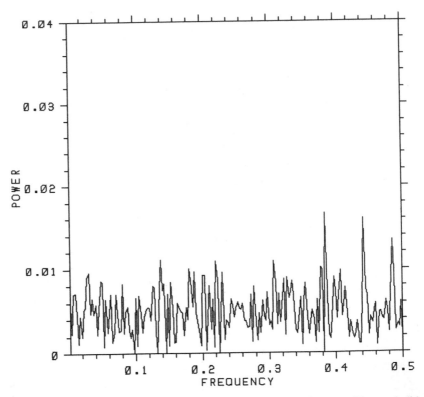

FIGURE 58.   (a) Power spectra of an observable from the logistic map with $\mu = 4$. (b) Power spectra of a white noise. The deterministic logistic evolution when it comes to the power spectra is indistinguishable from white noise.

This system is not autonomous since it is "externally" influenced. We can, however, transform it to an autonomous system by including a third equation:

$$\dot{x}_1 = x_2$$
$$\dot{x}_2 = -kx_2 - x_1^3 + b \cos t$$
$$\dot{t} = 1$$

Since t is now introduced as an independent variable, the system has effectively no forces acting on it from the outside. We now have an autonomous system of three variables. Thus, the phase space of Duffing's equation (and similar equations subjected to an external forcing) is three dimensional with coordinates (x_1, x_2, t) or, better yet, $(x_1, x_2, \cos t)$. In Fig. 60 the period-doubling scenario is illustrated for the dynamical system described

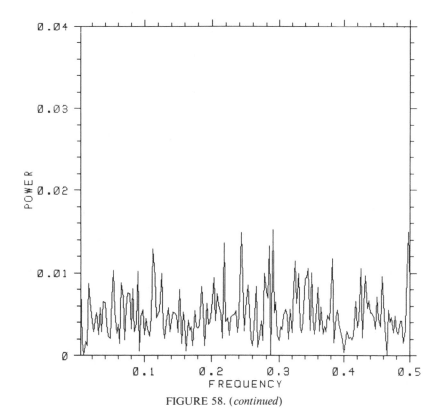

FIGURE 58. (*continued*)

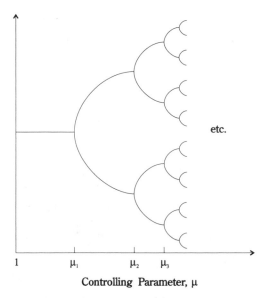

Controlling Parameter, μ

FIGURE 59.   Graphical representation of the periodic-doubling scenario.

by Eq. 6.16. Here the parameter b is kept a constant 0.05. The parameter k varies from (a) $k = 0.6$ (period 1) to (b) $k = 0.33$ (period 2) to (c) $k = 0.2845$ (period 4) to (d) $k = 7.5$ (chaos). The process by which the limit cycle gives rise to a period 2 orbit is called *flip* bifurcation. In analogy to Hopf bifurcation, during the flip bifurcation the limit cycle loses stability and gives birth to a cycle twice its period.

4. EXPLOSIVE BIFURCATIONS

Let us return again to the logistic map. Figure 61 shows the complete bifurcation diagram for the logistic map for $3 \le \mu \le 4$. This diagram presents a few surprises. The period-doubling scenario is evident for a while, but once chaotic evolution is achieved, further increases of μ result in stable periodic motions that appear as vertical windows in the bifurcation diagram. Most noticeable of all, the diagram depicts a period 3 limit cycle (indicated by the arrow near the right side). Within the period 3 window we see what appears to be three miniature copies of the entire diagram. Nearby (on the right) we observe chaotic behavior confined to three narrow islands that are always visited in the same sequence as the period 3 orbit.

The location of the chaotic orbit, however, within the bands is not ordered but seems erratic. This is often called noisy periodicity.[137] For slightly higher values of μ, however, the three narrow bands merge. In actuality, we see a bifurcation from a chaotic attractor in three narrow bands to a chaotic attractor over the entire interval. This is very nicely illustrated in Figs. 62a,b. Both figures show graphs of x_n versus x_{n+1} for $\mu = 3.855$ and $\mu = 3.9$, respectively. If the evolution is periodic of period 1, then on such a graph we observe just one point. If the evolution is of period 2, we observe two points, and so on. In Fig. 62, we observe three small "islands" that correspond to the chaotic evolution confined to the three narrow bands. In Fig. 62 we observe the "explosion" of the three narrow bands to an attractor occupying the entire interval. Such bifurcations cause a jump in the size of a chaotic attractor. They are often called explosive bifurcations and include "explosions" of periodic attractors to chaotic attractors, of chaotic attractors to larger chaotic attractors (like in our example), and of point attractors to periodic attractors.

These types of bifurcations are not like the period-doubling or Hopf bifurcations, which are called subtle bifurcations (see, for example, Abraham and Shaw.[4] Subtle bifurcations are those whose detection is possible only after the new behavior grows strong. We cannot detect the exact moment of bifurcation by casual observation. For example, let us assume that a period 1 cycle bifurcates to a period 2 cycle (see Fig. 63). The new cycle has twice the period and, hence, half the frequency. It stays close to the track of the period 1 cycle and goes around twice before closing. Thus, its second harmonic is equal to the fundamental frequency of the post–period 1 cycle and is quite strong. Thus, at the point of bifurcation one does not have a sharp indication that the split occurred. Only after the event can our "ear" clearly "hear" the difference.

5. FOLD BIFURCATION

Consider the simple system

$$\dot{x} = \mu - x^2 \tag{6.17}$$

where x is a scalar and μ is a controlling parameter. If $\mu < 0$, then the right-hand side is negative, and, as $t \to \infty$, $x \to -\infty$. If $\mu > 0$, the equilibrium points are given by $x^2 = \mu$. Thus, $\bar{x}_1 = \sqrt{\mu}$ and $\bar{x}_2 = -\sqrt{\mu}$. Equation (6.17) can be solved analytically:

$$\int \frac{dx}{\mu - x^2} = \int dt$$

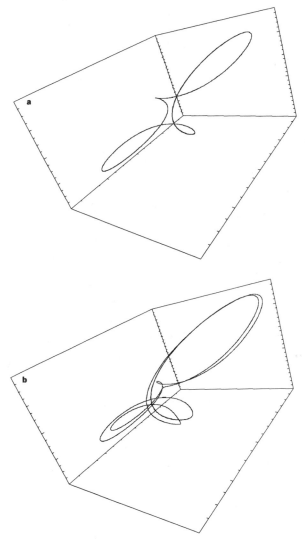

FIGURE 60.   Period doubling in the continuous system (flow) described by Eq. (6.16). The
parameter b is kept constant and equal to 0.05. The parameter k is varied: (a) $k = 0.6$ (period
1), (b) $k = 0.33$ (period 2), (c) $k = 0.2845$ (period 4), and (d) $k = 7.5$ (chaos).

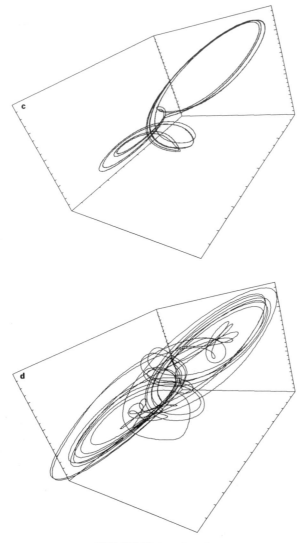

FIGURE 60. (*continued*)

Setting $\mu = a^2$ or $a = \pm\sqrt{\mu}$, we have

$$\frac{1}{2a}\ln\frac{a+x}{a-x} = t$$

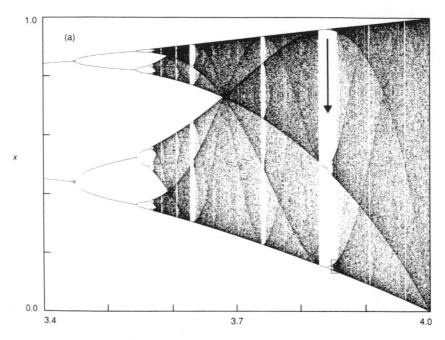

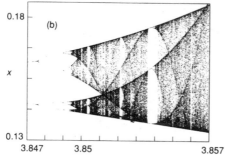

FIGURE 61. (a) The complete bifurcation diagram of the logistic map. The figure has been generated by plotting 300 values of the iterated function (after the transients have died out) for each of 1150 values of the control parameter μ. The map has a cycle of period 2 when the control parameter μ is at 3.4. This cycle "bifurcates" to cycles of periods 4, 8, 16, and so forth, as μ increase. Above $\mu \approx 3.57$ the map exhibits deterministic chaos interspersed with gaps where periodic motion has returned. For example, cycles of periods 6, 5, and 3 (shown by the arrow) can be seen in the three larger gaps to the right. (b) A blowup of the square in (a). The diagram displays properties of fractal structures. Magnifications of small parts resemble the whole structure. (Reproduced by permission from Roger Eckhardt, Los Alamos Laboratory. This figure appears in the book *From Cardinals to Chaos,* Cambridge University Press, 1989.)

or

$$x = \frac{\alpha(e^{2at} - 1)}{1 + e^{2at}}$$

For $a = \sqrt{\mu}$,

$$x = \frac{\sqrt{\mu}(e^{2\sqrt{\mu}t} - 1)}{1 + e^{2\sqrt{\mu}t}}$$

Thus, as $t \rightarrow \infty$, $x \rightarrow \sqrt{\mu} = \bar{x}_1$ (L'Hospital's rule applied here). For $a = -\sqrt{\mu}$,

$$x = \frac{-\sqrt{\mu}(e^{-2\sqrt{\mu}t} - 1)}{1 + e^{-2\sqrt{\mu}t}}$$

Thus, as $t \rightarrow \infty$, $x \rightarrow \sqrt{\mu} = \bar{x}_1$.

We therefore see that, for $\mu > 0$, there are two equilibrium points, one being an attractor (a stable point or node) and the other a saddle. Since $\bar{x}_1 - \bar{x}_2 = 2\sqrt{\mu}$, it follows that, as $\mu \rightarrow 0$, $\bar{x}_1 - \bar{x}_2 \rightarrow 0$. Therefore, as μ is lowered from positive to zero (see Fig. 64), the two equilibrium points come closer and closer to each other. Exactly at $\mu = 0$ the saddle and the node "collide" and annihilate each other. We say that the "path" of the attractor is interrupted. For negative μ values there is no attractor, and all solutions tend to $-\infty$.

Reversing the procedure, we then go from no attractor at all to the sudden creation of a saddle-node pair out of the blue. This type of bifurcation is called a static *fold*. The fold bifurcation can also occur in systems that exhibit cycles as their equilibria. In this case we have a saddle cycle, a stable attracting cycle, and an attracting fixed point (Fig. 65). The saddle cycle is a periodic repeller dividing the basin of attraction. All initial conditions inside the periodic repeller converge on the attracting cycle. As in the static fold, the attractor and the repeller approach each other and, at some point, annihilate each other. After this there are no limit cycles, and the whole plane is the basin of the point attractor, which was not involved. This type of bifurcation is often called a cyclic or periodic fold. Fold bifurcations fall into a new category of bifurcation, namely catastrophic. Unlike subtle bifurcations, catastrophic bifurcations are discontinuous (as indicated by the interrupted path of the attractor) and can thus be detected when they occur.

We have seen that there exist three major categories of bifurcations: subtle, catastrophic, and explosive. Some authors prefer to include an extra category, the fractal bifurcations that result from an infinite number of period doublings. We do not make this distinction here. We consider fractal bifurcations as a special case of subtle bifurcations. Figure 66 summarizes the various types of bifurcation discussed in this chapter.

A final note: We have categorized all the bifurcations discussed as subtle, catastrophic, or explosive, except the pitchfork. Pitchfork bifurcation can be subtle or catastrophic. In our example with the logistic equation there is no interruption of the attractor, it simply branches. Thus, the bifurcation is continuous and the pitchfork is subtle.

However, this is not always true, and annihilation of an attractor is involved. In these cases the pitchfork is catastrophic. Take, for example, the system $\dot{x} = \mu x + x^3$. The equilibrium points are $\bar{x}_1 = 0$, $\bar{x}_2 = \sqrt{-\mu}$, $\bar{x}_3 = -\sqrt{-\mu}$. The analytic solution of the system is

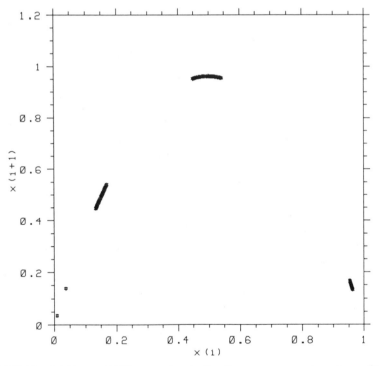

FIGURE 62.   (a) For $\mu = 3.855$ we see that the function x_{n+1} versus x_0 is made up of three small "islands." The islands are visited in the same sequence, but the location of the points inside each island is not ordered. (b) For $\mu = 3.9$ (just a bit greater than 3.855) the three small islands have "exploded" and now cover the entire interval. Such increases in attractor size are associated with explosive bifurcations.

$$x^2 = \frac{\sqrt{\mu}e^{2\mu t}}{1 - e^{2\mu t}}$$

If $\mu < 0$, then, as $t \to \infty$, $x \to \bar{x}_1 = 0$. For $\mu = 0$, $x = 0 = \bar{x}_1$. Thus, for $\mu \leq 0$ there is one attractor and two saddle points (i.e., all evolutions approach $\bar{x}_1$, not $\bar{x}_2$ and/or $\bar{x}_3$). For $\mu > 0$ the only physically meaningful equilibrium solution is $\bar{x}_1 = 0$, but as $t \to \infty$, $x \to \pm\sqrt{\mu}$. Thus, $\bar{x}_1 = 0$ is not an attractor. Therefore, we see that for $\mu \leq 0$ the two saddle points approach the attractor and collide with it at $\mu = 0$. The attractor is then annihilated, and for $\mu > 0$ we have only an unstable (saddle) point. In case this example does not clearly relate to the pitchfork, you may think of it in the opposite direction. As we go from positive to negative μ values, we end up with a pitchfork geometric configuration where attractors and saddles are interchanged (Fig. 67).

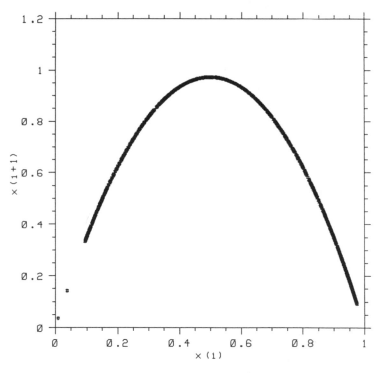

FIGURE 62. (*continued*)

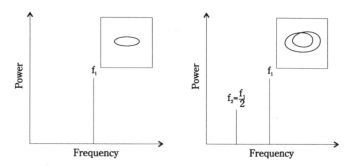

FIGURE 63.   Hopf and period-doubling bifurcations are called subtle bifurcations. As the diagram illustrates, after the bifurcation takes place the new cycle has twice the period, hence half the frequency. The new cycle stays close to the track of the period 1 cycle, but takes twice the time before it completes a closed loop. Therefore, even after the bifurcation, the period 2 orbit has memories of the path of the period 1 cycle. Because of that there cannot be a clear-cut point to indicate the exact time of the bifurcation. Thus, its second harmonic, which is equal to the frequency of the cycle before the bifurcation, is quite strong.

6. UNIVERSALITY AND ROUTES TO CHAOS

Recall the bifurcation diagram (Fig. 68) and define $\Delta\mu_i = \mu_{i+1} - \mu_i$. Also define ϵ_i as the scale of *branch splitting*. Then it can be shown[58] that

$$\frac{\Delta\mu_i}{\Delta\mu_{i+1}} \rightarrow \delta = 4.6692\cdots \qquad \text{for large } i$$

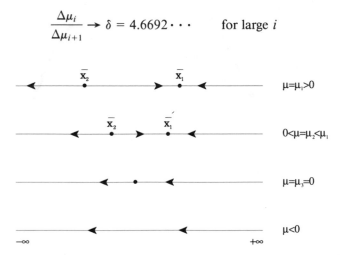

FIGURE 64.   Graphical representation of a fold bifurcation. Two initial stable equilibrium points approach each other as the controlling parameter changes until they collide and annihilate each other.

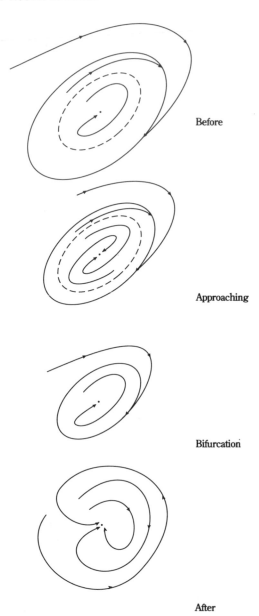

Before

Approaching

Bifurcation

After

FIGURE 65.   Graphical representation of a fold bifurcation where the system exhibits an attracting cycle and an attracting point and whose basin of attraction is divided by a saddle cycle (periodic repeller).

Bifurcation	Action	Characterization
Hopf		Subtle
Secondary Hopf		Subtle
Pitchfork	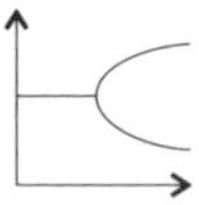	Subtle or Catastrophic
Flip		Subtle
Static Fold	 *nothing*	Catastrophic
Cyclic Fold	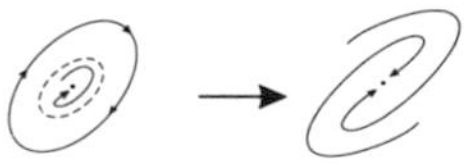	Catastrophic
Explosive	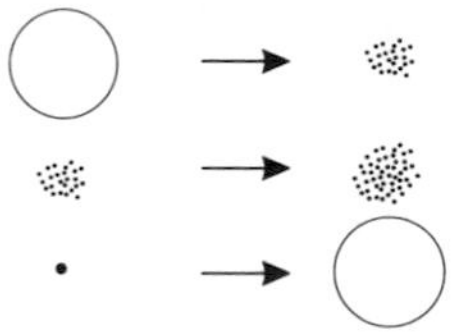	Catastrophic

FIGURE 66.   Summary of the various types of bifurcations discussed in this chapter.

and

$$\frac{\epsilon_i}{\epsilon_{i+1}} \to \alpha = 2.5029 \cdots \qquad \text{for large } i$$

The parameters δ and α are constants for any system that exhibits period doubling. Their beauty lies in the fact that, knowing δ, one can

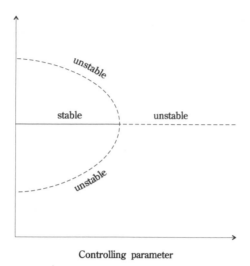

Controlling parameter

FIGURE 67.   Graphical representation of a situation where the pitchfork bifurcation involves annihilation of an attractor.

predict the value of the controlling parameter μ at which the next bifurcation occurs. The constant α controls the splitting of the trajectory in state-space. Note that, since the relative heights of successive subharmonics in the

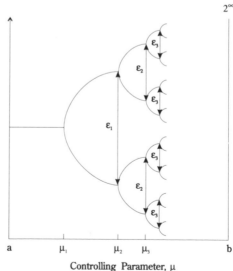

FIGURE 68.   In the bifurcation diagram for the logistic map, the branch splitting satisfies the relation $\epsilon_i / \epsilon_{i+1} = 2.5029$ for any i. Similarly $\Delta\mu_i / \Delta\mu_{i+1} = 4.6692$ for any i (where $\Delta\mu_i = \mu_{i+1} - \mu_i$).

power spectra measure this splitting, the spectra can also be used to obtain the parameter α. This certain quantitative behavior of a system is now referred to as *universality*. It is obvious from the definition of δ and α that $\Delta\mu_2 \rightarrow \Delta\mu_1/\delta$ and $\Delta\mu_3 \rightarrow \Delta\mu_2/\delta$. Thus, $\Delta\mu_3 \rightarrow \Delta\mu_1/\delta^2$. Generalizing: $\Delta\mu_n \rightarrow \Delta\mu_1/\delta^{n-1} \rightarrow 0$ as n becomes large. Similarly, for large n, $\epsilon_n \rightarrow 0$. Therefore, as $n \rightarrow \infty$, the destabilization of a stable orbit of period n occurs instantaneously. This continuous destabilization, which gives birth to a chaotic evolution, can, therefore, be thought of as an accelerating cascade process. Thus, this mechanism provides us with a possible "route" to chaos or a way to go from a simpler to a more complex behavior. Are there any other routes to chaos? In order to answer this question we must once again refer to the information given in Fig. 61.

In this figure we denote by $\mu^{(3)}$ the value of μ for which the period 3 orbit appears. For μ just below $\mu^{(3)}$ the orbit appears to be chaotic. We may, however, observe that the density of the points in the diagram appears to be greater near the impending period 3 orbit. Thus, the character of an orbit close to but below $\mu^{(3)}$ is as follows: It is periodic of period 3 for a long time; then it exhibits a short-term chaotic behavior ("intermittent bursts"). It then becomes periodic of period 3 again, and so on. As μ comes closer to $\mu^{(3)}$, the duration Δt of the period 3 behavior becomes larger and larger. In fact, $\Delta t \propto (\mu^{(3)} - \mu)^{-1/2}$ as $\mu \rightarrow \mu^{(3)}$.[92] Thus, at $\mu = \mu^{(3)}$, $\Delta t = \infty$ and a pure period 3 orbit appears. If we reverse the situation, we go from a period 3 orbit to a chaotic orbit. This presents a new route to chaos, *intermittency*.[169]

Figure 61 gives us the insight to discover yet another route to chaos. For a value of $\mu < \mu_c = 4$ there is a chaotic attractor. For $\mu > \mu_c$ there is no attractor. Thus, as μ is lowered from μ_c, a chaotic attractor appears. What happens at this point? For μ just greater than 4, any orbit initiating in the interval $0 < x < 1$ follows some erratic path for awhile. It then exits the interval $0 < x < 1$, becoming more and more negative in value and approaching $-\infty$. This transient is often referred to as a chaotic transient.[92] Its duration Δt depends on the initial condition. On the average (i.e., mean duration from many initial conditions) $\Delta t \propto 1/(\mu - \mu_c)^{1/2}$. For $\mu = \mu_c$, $\Delta t \rightarrow \infty$, meaning that if we reverse the procedure, that is, if we approach $\mu = \mu_c = 4$ from a value $\mu > 4$, the transient becomes of infinite length and itself becomes a chaotic attractor. This route to chaos is termed *crisis*. Both crisis and intermittency occur not only in maps like the one examined here but in many continuous systems (flows), and have been observed experimentally (see next chapter).

We have discussed three routes to chaos: period doubling, crisis, and intermittency. There is another. This route involves the destabilization of a torus with two frequencies (2-torus). According to Ruelle and Takens[186] and Newhouse et al.,[153] when a system that exhibits a 2-torus attractor makes the transition to a 3-torus attractor, the resulting topological attractor is unstable. With minimal perturbation it is converted to a chaotic attractor. Their results were initially challenged by experiments that provided evidence that 3-torus quasi-periodicity actually existed.[78,83,126] This disagreement was resolved following the work of Grebogi et al.[91] who examined the behavior of a dynamical system on a 3-torus that was perturbed by a nonlinear function with randomly chosen coefficients. They found that large perturbations sometimes produce a chaotic behavior, but the probability for producing chaos goes to zero as the amplitude of the perturbation becomes small. Thus, quasi-periodicity seems to be destroyed by small perturbations only if they are "chosen."

The transition from two-frequency quasi-periodicity to chaos is less understood than the period-doubling route to chaos. As explained in Ostlund et al.,[158] Aronson et al.,[8] and Feigenbaum et al.,[59] this transition is only universal when studied in a particular way. One has to use a two-parameter system in which both the ratio of the frequencies and the nonlinearity parameter are controlled. If the ratio is kept fixed at some irrational value, it should be possible to go directly to chaos as the nonlinearity parameter increases (universal way). If the ratio is not kept constant, then *phase locking* (where the evolution becomes periodic before it becomes chaotic) may take place.

7. THE GENESIS OF A STRANGE ATTRACTOR

Within its basin of attraction a dynamical system exhibits many different attractors. Depending on the controlling parameter, all sorts of periodic attractors might exist. When the system "changes" attractors, a periodic orbit becomes unstable while a new periodic orbit is born. This can go on until the system becomes chaotic. When the system is chaotic, the trajectory is thrown onto a nearby periodic orbit by its contracting eigenvalue (negative Lyapunov exponent). At the same time it is thrown out along the unstable direction toward another periodic orbit by its expanding eigenvalue (positive Lyapunov exponent). If the trajectory were to reach one of those periodic orbits, it would remain there forever, but when the system is chaotic this never happens. Instead, the trajectory wanders across the union of all unstable periodic orbits tracing a strange attractor. In a sense, the set of all periodic orbits can be viewed as a skeleton of the attractor, dictating its structure and thus its properties. For the Hénon map the number and stability of periodic orbits have been successfully related to its metric properties such as dimensions and Lyapunov exponents.

8. CHAOS: A SYNONYM OF RANDOMNESS AND BEAUTY

We discussed earlier that chaos is randomness generated by deterministic systems, and that chaotic evolutions are irregular, unpredictable evolutions exhibiting spectra practically indistinguishable from spectra of pure random processes. We also saw that a system can make the transformation from a regular periodic system to a chaotic system simply by altering one of the controlling parameters. Next we try to make the above a bit more "transparent." Consider again the logistic map

$$x_{n+1} = \mu x_n (1 - x_n)$$

For $\mu = 4$ we know that the equation generates chaotic evolutions. If we set $x = \sin^2 \pi y$, then we can write the equation as

$$\sin^2 \pi y_{n+1} = 4 \sin^2 \pi y_n (1 - \sin^2 \pi y_n)$$

or

$$\sin^2 \pi y_{n+1} = 4 \sin^2 \pi y_n \cos^2 \pi y_n$$

or

$$\sin^2 \pi y_{n+1} = \sin^2 2\pi y_n \qquad (6.18)$$

Equation (6.18) is equivalent to

$$y_{n+1} = 2 y_n \quad (\text{mod} 1) \qquad (6.19)$$

where (mod1) means drop the integer part of each y. Equation (6.19) is a linear difference equation, and it has a unique solution for each initial condition y_0. From Eq. (6.19) we have that

$$y_1 = 2y_0 \quad (\text{mod}1)$$

$$y_2 = 2y_1 = 4y_0 = 2^2 y_0 \quad (\text{mod}1)$$

$$y_3 = 2y_2 = 8y_0 = 2^3 y_0 \quad (\text{mod}1)$$

$$\vdots$$

$$y_n = 2^n y_0 \quad (\text{mod}1) \tag{6.20}$$

For simplicity, consider that Eq. (6.20) is $y_n = 10^n y_0 \,(\text{mod}1)$ instead of $y_n = 2^n y_0 \,(\text{mod}1)$. Then if

$$y_0 = 0.1234334678 \cdots \tag{6.21}$$

it follows that

$$y_1 = 10^1 \times 0.1234334678 \cdots \quad (\text{mod}1)$$

or

$$y_1 = 0.234334678 \cdots$$

Similarly,

$$y_2 = 0.34334678 \cdots$$

$$y_3 = 0.4334678 \cdots$$

and so on. Thus, we see that future iterates of $y_n = 10^n y_0 \,(\text{mod}1)$ may be obtained by moving the decimal point to the right of Eq. (6.21) and dropping the integer part. In our case $[y_n = 2^n y_0 \,(\text{mod}1)]$, we could write y_0 as a binary digit string

$$y_0 = 0.1110010101011000 \cdots \tag{6.22}$$

and obtain future iterates of Eq. (6.20) by moving the binary point to the right in Eq. (6.22) and dropping the integer part.

We can now see that the solutions of Eq. (6.20) become meaningful only if y_0 can be known exactly. In other words, the deterministic and unique solution of the logistic equation is unique only if y_0 is exactly known. If somewhere in Eq. (6.22) there is an approximation, it follows that all future iterates would be obtained with some error from their exact value.

From Eq. (6.22) we have that $\Delta y_n = 2^n \Delta y_0$. Thus, even small departures from some initial condition grow continuously. As a result, nearby trajectories diverge, and predictability in this case is limited. In fact, predictability would be limited even if somehow we could specify exactly y_0. If we go back to the logistic equation and assume that we know the initial condition exactly, say $y_0 = 0.9$, we find that

$$x_1 = 0.36$$

$$x_2 = 0.9216$$

$$x_3 = 0.2890136$$

and so on.

It follows that future iterates require more and more accuracy or information while their digit string soon becomes random. At some point the necessary information would be humanly impossible. Thus, at some point approximations take place, and predictability and complete determinism are compromised.

It is now easy to see that if the initial condition or some later iterate is a random string of zeros and ones (in a dyadic system) then all future iterates will be strings of randomly placed zeros and ones. Thus, we may say that Eq. (6.20) produces random numbers by passing the randomness of the initial value to the future values. In that sense chaos generates randomness or irregular evolutions exhibiting properties of random evolutions.

We should not, however, forget that the logistic equation displays all these "chaotic" consequences for $\mu = 4$. For $\mu = 2$ it is periodic of period 1. In this case if we again set $x = \sin^2 \pi y$, we have

$$\sin^2 \pi y_{n+1} = 2 \sin^2 \pi y_n (1 - \sin^2 \pi y_n)$$

or

$$\frac{1 - \cos 2\pi y_{n+1}}{2} = \frac{1}{2} \sin^2 2\pi y_n$$

or

$$1 - \cos2\pi y_{n+1} = 1 - \cos^2 2\pi y_n$$

or

$$\cos2\pi y_{n+1} = \cos^2 2\pi y_n \qquad (6.23)$$

An expression like Eq. (6.20) for Eq. (6.23) is rather intricate. We can see graphically, however, what happens in this case. Suppose $2\pi y_0$ corresponds to some angle as shown in Fig. 69. Then $\cos2\pi y_0 = a$ ($0 < a < 1$). It follows that $\cos^2 2\pi y_0 = a^2 < a$. Thus, if $\cos2\pi y_1 = a^2$, then $2\pi y_1$ should correspond to some angle whose cosine is equal to a^2. As shown in Fig. 69, there are two such angles. Let us consider the smaller angle and define the difference $2\pi y_1 - 2\pi y_0$ as ϕ_1 ($\phi_1 \propto a - a^2$). Repeating this procedure, we then find that $2\pi y_2$ should correspond to some angle whose cosine is equal to a^4 and that $2\pi y_2 - 2\pi y_1 = \phi_2 \propto a^2 - a^4$. Clearly, $\phi_2 < \phi_1$. In fact, as the iteration step $n \to \infty$, $\phi_n \to 0$. This is equivalent to saying that in this case evolutions from two nearby states will not diverge or that the dynamics will not be chaotic. Apparently for $\mu = 2$ the logistic equation does not pass the randomness of the initial condition's digit string. This proves that not all difference equations will generate chaos. It all

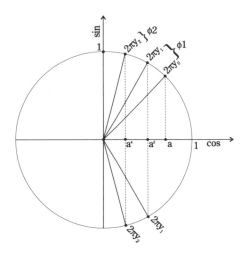

FIGURE 69.   Graphical proof of the nonchaotic nature of Eq. (6.23).

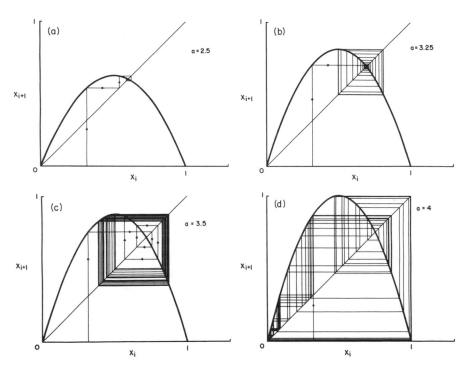

FIGURE 70. A given difference equation handles the iteration procedure uniquely. In the top (left) we have the logistic map for $\mu = 2.5$. Starting from some initial condition x_0, we obtain x_1, then x_2, and so on. Due to the unique shape of the curve for $\mu = 2.5$, the evolution soon converges to an equilibrium state of period 1. When $\mu = 4.0$ (bottom right), the curve is now different in shape and the iteration procedure does not converge to just one point. In fact, the shape of the curve causes the iteration sequence to become chaotic. In between we have evolutions of period 2 (top right) and period 4 (bottom left). (Figure courtesy of Dr. Leon Glass.)

depends on how a given difference equation manipulates its random x_n digit strings (see Fig. 70). This, in turn, leaves us wondering even more about the beauty of chaos.

CHAPTER 7

CHAOS ELSEWHERE

1. HAMILTONIAN CHAOS

We have documented the existence of chaos in dissipative systems. It is important to realize that chaotic behavior can arise in nondissipative Hamiltonian systems. Consider the following dynamical system known as the standard map:[34,131,226]

$$
\begin{aligned}
p_{n+1} &= p_n - \frac{k}{2\pi} \sin 2\pi q_n \\
q_{n+1} &= p_{n+1} + q_n
\end{aligned} \quad (\mathrm{mod}\, 1)
$$

$$(7.1)$$

where p, q, and n are the analogs of momentum, position, and time, respectively. The Jacobian J of system (7.1) is

$$
J = \begin{vmatrix} \dfrac{\partial f_1}{\partial p_n} & \dfrac{\partial f_1}{\partial q_n} \\[2mm] \dfrac{\partial f_2}{\partial p_n} & \dfrac{\partial f_2}{\partial q_n} \end{vmatrix}.
$$

where $f_1 = p_n - (k/2\pi) \sin 2\pi q_n$ and $f_2 = p_n - (k/2\pi) \sin 2\pi q_n + q_n$. Thus,

$$
J = \begin{vmatrix} 1 & -k \cos 2\pi q_n \\ 1 & 1 - k \cos 2\pi q_n \end{vmatrix} = 1
$$

which (since we are dealing with a map) means that volumes in phase space are conserved. Therefore, the system is conservative and there can be no attractors. The motion will not settle in some lower-dimensional subregion of the phase space. Due to the term $\sin q$, the motion is periodic in both q and p. The trajectory is given by the equation

$$q_n = p_0 + q_0 - \frac{k}{2\pi} \sum_{i=1}^{n} \sin q_i$$

where $k/2\pi$ is an irrational number. Thus, the combined motion is confined on a torus.

For $k = 0$ the system is clearly integrable:

$$p_n = p_0$$

$$q_n = np_0 + q_0$$

In this case p_n is constant, and q_n is increasing linearly in time. For $k \neq 0$ the system is nonintegrable. In this case the map is producing orbits that look like Fig. 71. Figure 71 is produced by generating orbits from many different initial conditions for $k = 1.1$. A periodic orbit in such a diagram would look like a closed loop of points. Figure 71 displays a striking, fine structure at all scales similar to those encountered in chaotic systems. At all scales one observes "islands" of closed loops, but one also observes "swarms" of dots that can only represent irregular nonperiodic motion. Figure 72 shows the full phase space of just one chaotic orbit for $k = 1.1$. If a point in the phase space is visited by the orbit, the point is "painted" black. Therefore, the white areas show forbitten regions. The interesting feature of this diagram is that the orbit contains holes at all scales. These holes represent regions in phase space reserved for periodic orbits, and thus they are not available for chaotic orbits.

The object in Fig. 72 displays properties appropriate for fractal sets. However, it is not a fractal set because it occupies a *finite* fraction of the phase space. As a result, the fractal dimension of the orbit is 2. Nevertheless, the object does contain holes at all scales, which means that the exact fraction of the area the orbit occupies would be a function of the resolution.

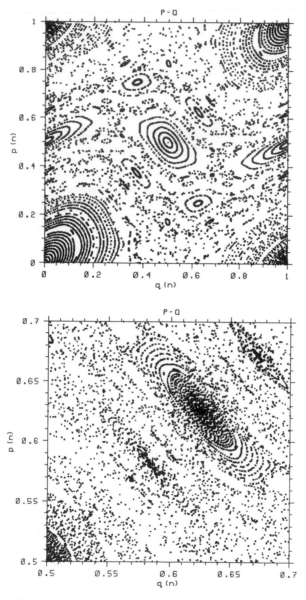

FIGURE 71.   The top figure shows orbits generated from the standard map for $k = 1.1$. Each orbit corresponds to some initial condition (q_0, p_0). A closed loop of points indicates a periodic motion. As we can see from the blown-up part in the bottom, islands of closed loops and swarms of dots are observed at all scales. The swarms of dots represent irregular nonperiodic motion.

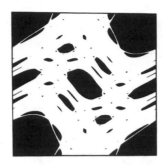

FIGURE 72.   The full phase space of *one* chaotic orbit for $k = 1.1$. If a point in phase space is visited, it is painted black. Thus, the white areas are forbidden regions. These holes are regions in phase space reserved for periodic orbits only. This figure displays properties appropriate for fractal sets but the dimension of this object is 2. Sets like that are called fat fractals. (Reproduced by permission from Dr. Doyne Farmer. This figure appears in the book *From Cardinals to Chaos,* Cambridge University Press, 1989.)

The approach to a finite value as the resolution goes to zero has definite scaling properties. Sets like this are called *fat fractal sets.*

There is a fundamental difference between the phase portraits just presented and those of dissipative systems. Because in conservative systems the area is conserved, there is no distinction between an attractor and a basin of attraction. The whole phase space is both the attractor and the basin, and, thus, any initial condition yields an orbit that eventually comes very close to the initial state. In dissipative systems the orbits wander away from the initial condition and go to the attractor. In dissipative systems the basin is much larger than the attractor. Remember that attractors approach some type of equilibrium and are thus associated with irreversibility and ergodic systems. In addition, all dissipative systems are nonintegrable. To explain the approach to equilibrium in nature we must, therefore, resort to dissipative systems. Furthermore, to explain the complexity and the nonperiodic character of most natural phenomena, we must resort to chaotic systems. Nevertheless, the chaotic behavior of some nonintegrable conservative dynamical systems suggests that there may be some connection between these systems and the "stochastic" behavior usually associated with irreversible dissipative systems.

2. QUANTUM CHAOS

We have established that Newtonian deterministic dynamics can be very random. In quantum mechanics randomness is contained in the wave function. Thus, quantum mechanics is nondeterministic to begin with. Is it possible that over and above this randomness there exists randomness surfaced by quantum chaos?

If we consider the case of a spatially bounded conservative quantum system with finite particle number, we can express the wave function $y(x, t)$ as

$$y(x, t) = \sum A_n(t) U_n(x) e^{-iE_n t/\hbar} \tag{7.2}$$

where $U_n(x)$ is the set of its energy eigenfunctions, x denotes the position variables, t denotes the time, and $A_n(t)$ denotes time-dependent coefficients. Substituting Eq. (7.2) into Schrödinger's equation

$$Hy = i\hbar \, \partial y/\partial t$$

we obtain

$$\frac{dA_n}{dt} + i\omega_n A_n = 0 \tag{7.3}$$

where $\omega_n = E_n/\hbar$ with E_n being an energy eigenvalue.

Following Ford [61] we may substitute $A(t) = e^{-i\theta(t)}$ [where $\theta(t)$ is an angle (mod 2π)] into Eq. (7.3) and obtain

$$\frac{d\theta}{dt} - \omega = 0 \tag{7.4}$$

We may now write Eq. (7.4) as

$$\theta(t + \Delta t) = \theta(t) + \omega \, \Delta t \quad (\text{mod} 2\pi)$$

or

$$\theta_{n+1} = \theta_n + \omega \quad (\mathrm{mod}2\pi)$$

or

$$\theta_n = \theta_0 + n\omega \quad (\mathrm{mod}2\pi) \tag{7.5}$$

Equation (7.5) indicates that θ_n increases linearly in time (in the same manner q_n increased in our example with the standard map for $k = 0.0$). No chaos here! There is simply no randomness in the Schrödinger equation. In fact, evidence suggests that even when we consider infinite, unbounded conservative quantum systems, a complete lack of randomness still seems to exist! This is an extraordinary and paradoxical conclusion that obviously raises many questions. One of the most pressing questions of quantum mechanics is its connection with the macroscópic (Newtonian) world. While research in the area of linking quantum physics and classical chaos is very infantile, recent results (Zhang et al.[227]) suggest that certain characteristics of a quantum system could be tied to chaotic behavior in the classical system that emerges from the quantum system.

3. LOW-DIMENSIONAL CHAOS IN INFINITE-DIMENSIONAL SYSTEMS

The scope of dynamical systems theory could be described as deterministic initial value problems, which include iterated mappings and initial value problems for ordinary differential equations (ODEs) and partial differential equations (PDEs). The existence of low-dimensional attractors in systems described by a finite number of ODEs is well established. The study of PDEs under the prism of chaos might seem hopeless given the fact that the phase space of a PDE system is infinite. Nevertheless, in many instances geometric ideas can be successfully applied to PDEs, especially where dissipation plays an important role. If the effect is strong, then the contraction of volumes in phase space might force the system to settle to a low-dimensional attractor. Such a system, which has attracted considerable attention, is the Kuramoto-Sivashinsky (K-S) equation

$$\frac{\partial u}{\partial t} + u\frac{\partial u}{\partial x} + \frac{\partial^2 u}{\partial x^2} + v\frac{\partial^4 u}{\partial x^4} = 0$$

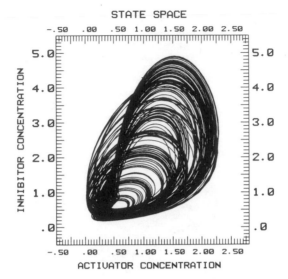

STATE SPACE

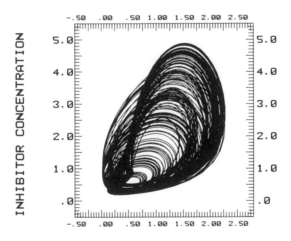

ACTIVATOR CONCENTRATION

FIGURE 73. Results of a reaction-diffusion model for biological pattern formation consisting of a row of 100 cells. The orbits of cell 10 (top) and cell 40 (bottom) are shown. Since the initial conditions are slightly different, the trajectories are different. The shape of the attractor, however, is the same. Each cell is, thus, undergoing a chaotic evolution on what appears as a common attractor. This results in the formation of spatial gradients in the resulted from the evolution of the one-dimensional cell array pattern. Also in such cases the original phase space of 102 dimensions may settle down to a much lower dimensional space. (Tsonis *et al.*[216])

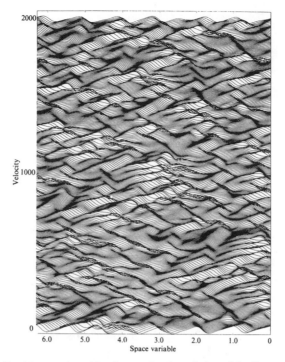

FIGURE 74.   Spatial pattern resulting from a chaotic evolution of the Kuramoto-Sivashinsky equation. The generation of spatial inhomogeneities and its relation to chaotic evolution in a system of PDEs is an important characteristic of the dynamic of such a system. (Reproduced by permission from Frisch *et al.*[71].)

where u indicates velocity and x is a space variable. In studies by Kuramoto and Tzuzuki,[121] Kuramoto,[120] Frisch *et al.*,[71] Chaté and Manneville,[32] and Hyman and Nicolaenko,[110] it has been established that the K-S system behaves as a finite-dimensional system of ODEs. The solution to the K-S system reveals a complex interplay between simple spatial patterns and low-dimensional chaos, thus bridging the gap between infinite-dimensional behavior of PDEs and finite-dimensional behavior of ODEs.

Another system that seems to behave like a low-dimensional system is the forced reaction-diffusion (R-D) model for biological pattern formation (Tsonis *et al.*[216]). The mathematical formulation of the R-D system is

$$\frac{\partial a}{\partial t} = \frac{c(a^2 + c_0)}{h} - \mu a + D_a \nabla^2 a + A \sin\omega t$$

$$\frac{\partial h}{\partial t} = ca^2 - \nu h + D_h \nabla^2 h$$

where t is the time, a is the activator concentration, h is the inhibitor concentration, D_a is the rate at which the activator diffuses from cell to cell, D_h is the rate at which the inhibitor diffuses from cell to cell, μ is the decay rate of the activator, ν is the decay rate of the inhibitor, c is the source density, and c_0 is an activator-independent activator production. The term $A \sin\omega t$ represents external forcing. The parameter A is the amplitude of the forcing, and ω is the angular frequency ($2\pi/p$, where p is the periodicity).

Given a row of n adjacent cells, the solution of the coupled nonlinear PDEs provides the evolution of the cells in time from a slightly perturbed, unstable steady state. According to this formulation, one reaction is autocatalytic and the other acts antagonistically to the autocatalysis. Any vital molecule for the generation of a pattern is the activator a, which stimulates its own production (autocatalysis). The antagonistic reaction can be caused by the inhibitor h. As is demonstrated in Tsonis *et al.*[216] for some choice of the controlling parameters, the evolution of a and h for each cell is chaotic. For a row of 100 adjacent cells, Fig. 73 shows the state-space of cell 10 (top) and cell 40 (bottom). Note that the two trajectories are somewhat different. This is to be expected, since a chaotic system will produce different evolutions starting from two slightly different initial conditions. What is interesting, however, is that the attractor seems to be the same for both cells (in fact it seems to be the same for all cells). In such a case a system of an original phase space of 102 dimensions (100 cells plus h and a) may settle down to a much lower dimensional space.

Due to different evolutions, patterns generated from one-dimensional arrays of cells exhibit spatial gradients. In such cases very interesting spatial patterns may form. An example is presented in Fig. 74, which shows the chaotic evolution of strongly perturbed cellular solution of the K-S system (for more examples see Ciliberto and Rubio[35]).

PART III

APPLICATIONS

The knowledge at which geometry aims is the knowledge of the eternal.
—PLATO, REPUBLIC, VII, 527

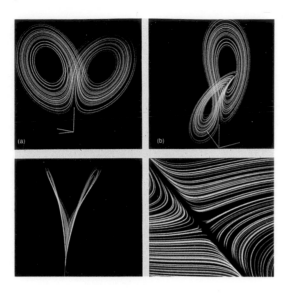

FIGURE 29.   (Reproduced with permission from Dr. Gottfried Mayers-Kress, Los Alamos National Laboratory. This figure appears in the book *From Cardinals to Chaos,* Cambridge University Press, 1989.)

FIGURE 35.   (Reproduced with permission from Dr. C. Pickover, IBM. This figure appears in his book *Computers Pattern, Chaos, and Beauty,* St. Martin Press, 1990.)

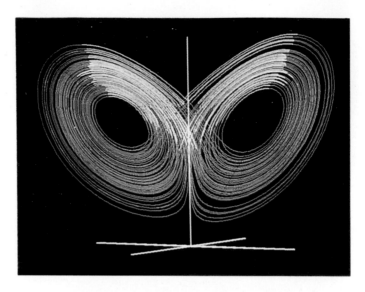

FIGURE 44. (Figure courtesy of Dr. J. Nese.)

FIGURE 47. (Figure courtesy of Dr. C. Grebogi, University of Maryland. This figure appears in *Science* **238**, 1987, copyright 1987 by the AAAS.)

FIGURE 104. (Reproduced by permission from Ottino.[160])

FIGURE 107. (Reproduced by permission from Dr. W. J. Freeman, 1991.)

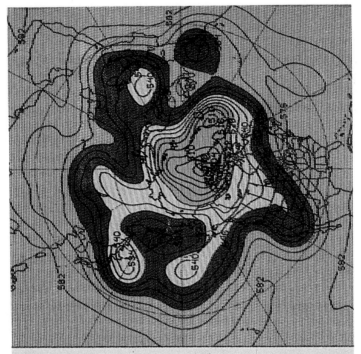

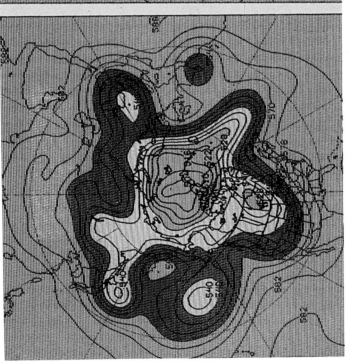

FIGURE 128. (Reproduced by permission from Dr. Kalnay of the National Meteorological Center and from *Weatherwise*, Heldref Publications.)

CHAPTER 8

Reconstruction of Dynamics from Observables

1. PHASE-SPACE RECONSTRUCTION—METHOD OF DELAYS

The study of the mathematical dynamical systems presented in the previous chapters advanced our understanding of the dynamics of nonlinear deterministic systems. We now know that random-looking behavior can arise from simple nonlinear systems. Such dynamics, now termed chaotic dynamics, exhibit complicated strange attractors that are fractal sets with positive Lyapunov exponents. We also learned how the dynamic behavior of a system can change via bifurcations, and how period doubling, intermittency, and crisis can take a system from a periodic to a nonperiodic evolution.

If the mathematical formulation of the system is given, recognizing chaotic behavior is as easy as producing the Fourier spectra of the evolution of one of the variables. Since the evolution is deterministic, broadband noise spectra would be sufficient to identify chaos. Furthermore, since the number of variables is known, the generation of the state-space and the attractor, as well as the estimation of the various dimensions and Lyapunov exponents, is straightforward. However, when we deal with controlled experiments where we cannot record all the variables, and/or with observables from some uncontrolled system (like the atmosphere) whose mathematical formulation and total number of variables may not be known exactly, life becomes a little bit more complicated. Fourier analysis alone cannot be used for proof of chaos since the observable might be a random variable. Thus, more evidence, such as dimensions, Lyapunov exponents, and phase-space trajectories, must be provided. To acomplish this, we must first have a way to reconstruct the phase space of the underlying dynamical system (if any) from observables.

The phase space can be approximated by using a single record of some observable $x(t)$ according to a procedure outlined in Packard et al.,[162] Ruelle,[183] and Takens.[203] This procedure calls for the generation of the complete state vector $X(t)$ by using $x(t)$ as the first coordinate, $x(t + \tau)$ as the second coordinate, and $x(t + (n - 1)\tau)$ as the last coordinate, where τ is a suitable delay parameter and n is the *embedding* dimension.

A few words about the definition of the embedding dimension are in order at this point. When an attractor exists, its dimension is smaller than that of the dimension of the state-space. We could take advantage of this by trying to develop a lower-dimensional dynamical system that describes only the motion on the attractor. This can be achieved by embedding the attractor in a smooth manifold (smooth in the sense that it does not possess self-intersections) and restricting the model to this manifold. Recall that a manifold is a geometric model (usually a coordinate space, but it could be a cylinder, a torus, etc.) used to completely describe a phenomenon. The lowest possible dimension of such a manifold is called the embedding dimension. Attractors that are topological structures (points, limit cycles, tori) are submanifolds of the manifold in which they are embedded. Attractors that are fractal sets are *not* submanifolds (fractals are not manifolds). When we try to reconstruct the attractor from an observable, the dimensionality of the manifold that contains it is not known *a priori*. Thus, the embedding dimension is varied until we "tune" to a structure that becomes invariant (more on this will follow). According to Whitney's theorem, any smooth manifold of dimension m can be smoothly embedded in $n = 2m + 1$ dimensions. In addition, Takens[203] showed with regard to reconstructions that if the dimension of the manifold containing the underlying attractor is m, then embedding the data in a dimension $n \geq 2m + 1$ preserves the topological properties of the attractor. More specifically, the embedding will be a diffeomorphism—a differentiable mapping with a differentiable inverse—from the true phase space to the delay space. Thus, reconstructions preserve geometrical invariants such as dimension, positive Lyapunov exponents, etc. Note that, strictly speaking, Whitney's theorem is meaningful when an infinite data set is available. When we are dealing with finite data sets, we may be looking at local characteristics of the attractor. In such cases, the word "embedding" is used loosely as any topologist will point out. Table I shows the time delay procedure as applied to a hypothetical times series of size $N = 10$. Note that for an embedding dimension n there remain $N_n = N - (n - 1)\tau/\Delta t$ points.

Figure 75a shows the Rössler attractor projected onto the xy plane.

TABLE 1
Illustration of the Time Delay Procedure of Phase-Space
Reconstruction from an Observable $x(t)$

Time step t	1	2	3	4	5	6	7	8	9	10
Observable $x(t)$	3	3	4	2	1	1	0	6	7	2
First coordinate										
$\quad x(t)$	3	3	4	2	1	1	0	6	7	2
Second coordinate										
$\quad x(t + \tau)$	3	4	2	1	1	0	6	7	2	
Third coordinate										
$\quad x(t + 2\tau)\, \tau = 1$	4	2	1	1	0	6	7	2		

The attractor has been generated by numerical integration of the Rössler system [Eq. (5.3)]. The trajectory tends to a strange attractor possessing a fractal (Cantor-like) structure and a chaotic fold.

Figure 75b shows the reconstructed attractor using the method of delays with $\tau \sim 0.017$ mean orbital periods. This figure is a projection of a 3D embedding in two dimensions. We observe an extreme concentration of states along the diagonal that masks the attractor structure. The problem here is that a very small τ was chosen. When τ is very small, then $x(t + \tau)$ $\sim x(t)$ and thus all trajectories appear to lie on the line $x(t + \tau) = x(t)$. In such cases the dimension of the attractor is close to 1, which is smaller than the actual dimension of 2.07.

Figure 75c shows the recovered attractor for $\tau = 0.23$ mean orbital periods corresponding to the first zero of the autocorrelation function. This is also a projection of a 3D embedding in two dimensions. Now the reconstruction has preserved the fractal nature of the attractor as well as the chaotic fold. This example demonstrates that the method of delays is not without problems. A proper delay time must be chosen.

2. THE CHOICE OF τ

Every time we have to perform a statistical analysis on a certain sample, we must make sure that the analysis involves independent values. This must be so because dependent values bias our estimations. The same philosophy applies when we reconstruct the attractor by producing a cloud of points at a given embedding dimension. If points that are not independent to previously generated points are included, the estimation of the correlation

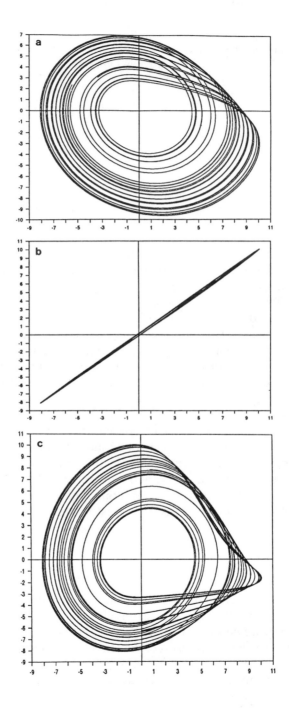

or any other dimension may be biased (typically underestimated). There-fore, τ must be chosen so as to result in points that are not correlated to previously generated points. Thus, a first choice of τ should be in terms of the decorrelation time of the time series under investigation.

The question now arises: how do we define the decorrelation time? A straightforward procedure is to consider the decorrelation time equal to the lag at which the autocorrelation function first attains the value zero. Other approaches consider the lag at which the autocorrelation function attains a certain value, say $1/e$, 0.5,[191] or 0.1.[210] Another suggestion is $\tau = T/n$, where T is the dominant periodicity (as revealed by Fourier analysis) and n is the embedding dimension.[225] In this way τ gives some measure of statistical independence of the data averaged over an orbit and thus is an appropriate approach if the autocorrelation function is periodic. Another suggestion[162] is that τ should satisfy $\tau \ll I/\Lambda$, where I is the precision of measurement and Λ is the sum of all positive Lyapunov ex-ponents of the flow. This ensures that the information-generating properties of the flow do not randomize information between successive sites on the recorded attractor. This approach, however, is not practical when we are dealing with an observable from an unknown dynamical system whose Lyapunov exponents are what we seek. As pointed out by Frazer and Swin-ney,[68] however, the autocorrelation function measures the linear depen-dence between successive points and may not be appropriate when we are dealing with nonlinear dynamics. They argue that what should be used as τ is the local minimum of the mutual information, which measures the general dependence between successive points.

The basic philosophy behind the definition of mutual information is as follows: Given an N-element sequence, we may calculate the transition probabilities $P_s(s_i)$ that a measurement of the state s yields s_i. The infor-mation entropy is then defined as

←

FIGURE 75.   (a) The Rössler attractor projected on the xy plane. The trajectory has been generated by numerical integration of the Rössler system [Eq. (5.3)] for $a = 0.2$, $b = 0.4$, and $c = 5.7$. (b) The Rössler attractor reconstructed from the variable $x(t)$ using a delay time $\tau \approx 0.017$ of the mean orbital period. This figure is a projection of a 3D reconstruction [$x_1 = x(t)$, $x_2 = x(t + \tau)$, $x_3 = x(t + 2\tau)$] in two dimensions (x_1, x_2). Because τ is very small, $x(t + \tau) \approx x(t)$ and, thus, all points tend to fall on the diagonal. (c) As in (b), but for $t \approx 0.23$. This τ corresponds to the lag at which the autocorrelation function attains for the first time a zero value. In this case the major characteristics of the attractor are recovered. (Figure courtesy of Dr. G. W. Frank.)

$$H(s) = -\sum_{i=1}^{N} P_s(s_i)\log P_s(s_i) \qquad (8.1)$$

To understand the usefulness of Eq. (8.1) it is helpful to think of entropy as a measure of the uncertainty associated with the measurement of state s or the "quantity of surprise you feel when you read the result of a measurement." Low-probability (unexpected) measurements carry greater entropy than do high-probability measurements (zero probability $\rightarrow$ complete surprise $\rightarrow$ infinite entropy; probability 1 $\rightarrow$ no surprise at all $\rightarrow$ zero entropy).

We can now ask how $x(t + \tau)$ depends on $x(t)$ as a function of τ. Accordingly, by making the assignment $s = x(t)$, $q = x(t + \tau)$, we may define the conditional entropy:

$$H(q, s_i) = -\sum_{j=1}^{N} \left[\frac{P_{sq}(s_i, q_j)}{P_s(s_i)}\right]\log\left[\frac{P_{sq}(s_i, q_j)}{P_s(s_i)}\right]$$

where $P_{sq}(s_i, q_j)$ is the probability that measurements of s and q yield s_i, q_j. In this case $H(q, s_i)$ is the uncertainty of q, given s_i. The mutual information is then defined to be the amount that a measurement of $s = s_i$ reduces the uncertainty of q:

$$I(q, s_i) = H(s_i) + H(q) - H(s_i, q)$$

If the delay time τ is chosen to coincide with the first minimum of the mutual information, then the recovered state vector X will consist of components that possess minimal mutual information between them. Figure 76 gives an example of attractor recovery using as τ (1) the lag at which the autocorrelation attains the value of zero for the first time (left) and (2) the minimum of mutual information (right). Note the difference in the quality of the delineated portraits. Those differences might be important when we want to accurately estimate the dimension and the Lyapunov exponents of the corresponding dynamical system.

The mutual information method is definitely the most comprehensive method of determining proper delay times. The only problem associated with this method is that it requires a large amount of data and that it is computationally cumbersome (this, however, should not constitute an excuse for not using it). A somewhat related approach that does not demand as much data has been recently proposed by Liebert and Schuster.[132]

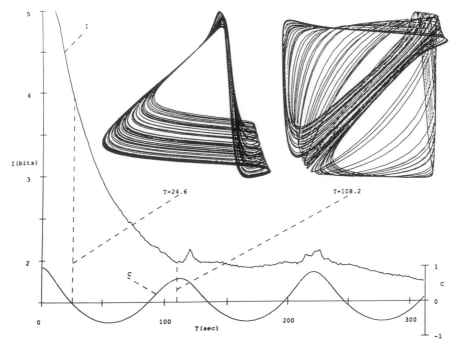

FIGURE 76. Phase portraits of the attractor in the Belousov-Zhabotinsky reaction (see Chapter 9 for details on this reaction). The autocorrelation C and mutual information I as a function of T are also included. On the left the attractor recovery using as τ the lag at which the autocorrelation function attains for the first time a value of zero is shown. On the right the reconstructed attractor using as τ the first minimum of the mutual information is shown. Differences can be observed. Those differences can be important, especially when we wish to estimate various measures from observables. (Reproduced by permission from Frazer and Swinney.[68])

According to this approach a good choice of τ is the first minimum of the generalized correlation integral $C(\tau, r, n)$. For some r and embedding dimension n, one can calculate the correlation function $C(r)$ (defined in Chapter 5) as a function of τ. The resulting function gives the generalized correlation integral. The logarithm of $C(\tau, r, n)$ is a measure of the averaged information content in the reconstructed vectors, and thus its minimum provides an easy way to define a proper τ.

For some attractors, it really does not matter if the autocorrelation function or the mutual information or the correlation integral is used. For example, for the Rössler system all approaches suggest a $\tau \sim \frac{1}{4}$ of the mean orbital period. For other attractors, however (such as the one in Fig. 76), the estimation of τ might depend strongly on the approach employed.

We recommend that the mutual information or the correlation integral be used to define a proper τ. If we have to rely on the autocorrelation function, then the results should be checked for a variety of τ's just to make sure they do not change significantly.

When applicable, there is a way to avoid having to define a proper delay time. This way requires not one but n observables of the same variable. Each sequence can be sampled independently of the other and thus could be used as an independent coordinate in an n-dimensional phase space. For example, instead of dealing with one time series representing atmospheric pressure at a given point, we may measure pressure at n independent points, thus obtaining n such time series. Instead of sifting one time series to obtain the phase space coordinates, we simply bring in one new time series at a time.

3. PHASE-SPACE RECONSTRUCTION—SINGULAR SYSTEM APPROACH

Another method of phase-space reconstruction is based on the singular system approach.[26] This approach considers lagged copies of an observable $x(t)$ sampled at equal intervals Δt, $x_i = x(t_0 + M \Delta t)$, $1 \leq M \leq N$, and estimates the eigenvalues λ_n and eigenvectors ρ_n of their covariance matrix C ($1 \leq n \leq M < N$). The eigenvectors are often called empirical orthogonal functions (EOFs). In this case the covariance matrix has the form

$$C_{ij} = \frac{1}{N_n} \sum_{l=1}^{N_n} x(t_l + i\tau)x(t_l + j\tau), \qquad i, j = 0, \ldots, n - 1$$

where $N_n = N - (n - 1)\tau/\Delta t$.

Orthogonality in the embedding space is related to the statistical properties of the time series. More precisely, EOFs being the eigenvectors of the covariance matrix, which is a matrix of quadratic averages (average here refers to a mean over all data points), they describe variables that are (statistically) linearly independent.

The space spanned by the eigenvectors is called the singular space. Embedding the data in this singular space amounts to switching from the state vector X to a state vector $Y = XC$, where C is an $n \times n$ matrix whose columns are the eigenvectors ρ_n. Eliminating off-diagonal terms of the covariance matrix provides an n-dimensional coordinate system in which the variables possess the greatest average degree of statistical independence. Such a system is called a Karhunen-Loeve coordinate system, [26] and viewing the recovered trajectory from this system minimizes the effect of arbitrary delay times on the reconstruction.

Figure 77 shows the Rössler attractor recovered from a 3D embedding, projected to a 2D Karhunen-Loeve coordinate system. Note the similarity to Fig. 75c. Unfortunately this approach is not without problems. The statistical independence of the variables is linear. This might not be the desired result, since it is the *nonlinearity* of the system that is important. In addition, since the dimension of the phase space in which an attractor should be embedded is not known *a priori,* estimating the dimension of an attractor from just one observable requires reconstructions in successively higher embedding dimensions (more details are given in the following section). One hopes that for higher than required embedding dimensions the recovered trajectory is "nested" in a lower dimensional subspace.

Figure 78 shows the projection of a five-dimensional reconstruction of the Rössler trajectory in two dimensions. Since the actual dimension of the Rössler attractor is ~2.07, one would expect the trajectory to lie in a 3D subspace of the five-dimensional embedding space. In this case this

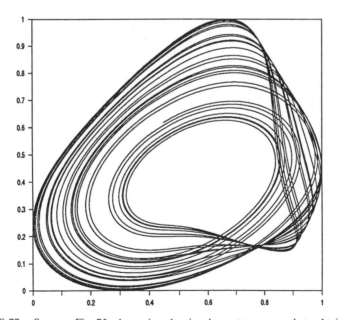

FIGURE 77.   Same as Fig. 75c, but using the singular system approach to obtain the coordinates of the embedding space. (Figure courtesy of Dr. G. W. Frank.)

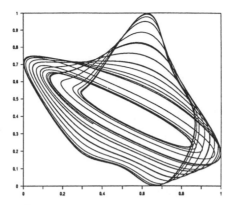

FIGURE 78.   Same as Fig. 77, but now we project a 5D reconstruction using the singular system approach in two dimensions. The differences between this figure and Fig. 77 occur because the recovered trajectory does not lie within a 3D subspace but extends along each of the five available degrees of freedom (dimensions). (Figure courtesy of Dr. G. W. Frank.)

projection should be similar to the one in Fig. 77. The recovered trajectory in Fig. 78 is quite similar to the recovered trajectory in Fig. 77, but we definitely see certain distortion in the structure of the attractor. As the eigenvalues of the covariance matrix C indicate, the distortions arise because the trajectory does not lie within a 3D subspace, but instead extends along each of the 5 degrees of freedom available to it. In fact, Mees et al.[147] have shown that this expected "nesting" of the trajectory is not a generic quality of such reconstructions. This limits the usefulness of the singular space method as a dimension estimator.

Frazer[67] demonstrated that the method of delays with τ's determined from mutual information is superior to the singular system approach. The superiority arises from the notion of general independence between points (which mutual information provides), as opposed to the notion of linear independence (which the singular system approach provides).

In general, we would suggest using the method of delays when a proper τ can be defined. In any case we recommend using both approaches and comparing the results. If the results are similar, the evidence becomes stronger and more convincing.

4. ESTIMATING DIMENSIONS

Assuming that an appropriate phase-space reconstruction is at hand, we can now estimate the dimension(s) of the underlying attractor (if any).

For an n-dimensional phase space, a "cloud" or set of points is generated. The capacity dimension of this set can be estimated by covering the set by n-dimensional hypercubes of side length l and determining the number, $N(l)$, of hypercubes needed to cover the set in the limit as l goes to zero.[141] This is the box-counting algorithm, and if $N(l)$ scales as

$$N(l) \propto l^{-d} \qquad l \to 0 \tag{8.2}$$

then the scaling exponent d is an estimate of the capacity dimension for that n.

The capacity dimension is also referred to as the fractal or Hausdorff-Besicovitch dimension. In general, the capacity dimension and the Hausdorff-Besicovitch dimension are often considered equivalent, but there is a fine distinction between them. The Hausdorff-Besicovitch dimension is obtained from covering the set minimally with hypercubes that may be different in size. The capacity dimension involves the same process except that the size of the hypercubes is the same.[50] In a $\log N(l)$ versus $\log l$ plot, the exponent d can be estimated by the slope of a straight line (the scaling region). Using the state vector $X(t)$, we can test Eq. (8.2) for increasing values of n. If the original time series is random, then $d = n$ for any n (a qualified random process of infinite sample size embedded in an n-dimensional space always fills that space). If, however, the value of d becomes independent of n (that is, reaches a saturation value D_0), it means that the system represented by the time series has some structure and should possess an attractor whose capacity dimension is equal to D_0.

The foregoing procedure for estimating D_0 is a consequence of the fact that the actual number of variables in the evolution of the system is not known; thus, we do not know *a priori* what n should be. We must, therefore, vary n until we "tune" to a structure that becomes invariant in higher embedding dimensions (an indication that extra variables are not needed to explain the dynamics of the system in question). As will become clearer, however, this condition is necessary (for noiseless data sets) but may not be sufficient.

This numerical approach to estimating the dimension of an attractor from a time series is limited, however. The reason is that an enormous number of points on the attractor is required to make sure that a given area in the phase space is empty and not simply rarely visited. It has been documented [73,93] that a box-counting approach is not feasible for phase-space dimensions greater than 2.

An alternative approach, which is more applicable, has been developed by Grassberger and Procaccia.[88,89] This approach again generates a cloud of points in an n-dimensional phase space. But instead of covering the set with hypercubes, it finds the number of pairs $N(r, n)$ with distances less than a distance r. In this case, if for significantly small r, we find that

$$N(r, n) \propto r^{d_2} \qquad (8.3)$$

then the scaling exponent d_2 is the correlation dimension of the attractor for that n. We then test Eq. (8.3) for increasing values of n and check, as previously, for a saturation value D_2 that estimates the correlation dimension of the attractor. The correlation dimension D_2 is less than the capacity (or fractal) dimension D_0 and is actually a measure of the extent to which the presence of a data point affects the position of the other points lying on the attractor. For a random time series there is no such "spatial correlation" in any embedding dimension, and thus no saturation is observed in d_2.

This approach still requires numerous points (especially for high embedding dimensions), but at least it is more feasible than the box-counting method, because a covering by small hypercubes becomes virtually impossible for embedding dimensions greater than or equal to 3.[87,88] Thus, most analyses to date have concentrated on calculating D_2.

Note that one important consequence of knowing the dimension of an attractor is that the dimensionality of an attractor, whether fractal or not, gives information on the minimum number of variables present in the evolution of the corresponding dynamical system. In other words, the attractor must be embedded in a state-space of at least the smallest integer larger than its dimension. Therefore, the determination of the capacity dimension (or for that matter of any other generalized dimension) of an attractor sets a number of constraints that should be satisfied by a model used to simulate the evolution of the system.

Figure 79 outlines the procedure of estimating the correlation dimension of a free-of-noise infinite data set. We start with embedding the data in two dimensions. This generates a cloud of points (Fig. 79a). Then we calculate $C(r)$ as a function of r. We then plot $\log C(r)$ versus $\log r$ (Fig. 79b). This function is expected to look like Fig. 79b because as r approaches the diameter of the attractor there are fewer and fewer pairs whose separation is as large as the diameter. Thus, for values of r greater than the size of the attractor, no pair is found, and the number of pairs with distances less than r remains the same as r increases (the edge effect).

Having produced Fig. 79b, we want to find the scaling region (if any).

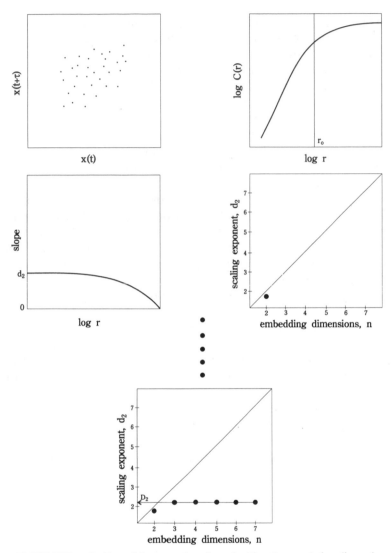

FIGURE 79.   Outline of the procedure for estimating the correlation dimension.

The best way to do this is to produce a figure such as Fig. 79c, which shows the slope of the function $\log C(r)$ as a function of $\log r$. The slope of a straight line is constant. Therefore, if a scaling region exists [$\log C(r) = a + b \log r$], then in a slope versus $\log r$ graph one should observe a plateau.

This plateau provides an estimate for the correlation dimension for $n = 2$. Having the estimate, we then locate a point in a d_2 versus n graph (Fig. 79d). By the way, the worst procedure to find the scaling region is "eyeballing." Avoid it! We then repeat the procedure for successively higher embedding dimensions n. If at some point d_2 becomes independent of n (i.e., it reaches a saturation value D_2), this indicates an underlying attractor of dimension D_2 (Fig. 79e).

Direct application of the Grassberger-Procaccia algorithm requires computer time that scales as $O(N^2)$. That can make the algorithm computationally cumbersome, especially when we are dealing with (or require) large sample sizes.

An efficient algorithm for estimating the correlation dimension has been developed by Theiler. [205] The philosophy behind Theiler's improvement is that it is not necessary to include in the calculations distances greater than a cutoff distance r_0 (see Fig. 79b), since they do not really define the scaling region. Theiler's algorithm (often referred to as the box-assisted algorithm) requires computer times that scale as $O(N \log N)$. Thus, depending on N, speedup factors up to 1000 over the usual method can be achieved. A further improvement was suggested by Grassberger, [86] who proposed an optimized box-assisted algorithm that reduces the computer time by an additional factor of 2–4. Recently, Liebovitch and Toth [133] presented a fast algorithm to compute the capacity dimension of low-dimensional dynamic and iterated (maps) systems. The corresponding codes are available from the authors.

Because it is relatively easy to implement, the Grassberger-Procaccia algorithm has received a lot of attention and been applied in many studies. One of the major problems, however, in many of those studies is the number of points used in the calculations. This is an important issue and deserves an in-depth discussion.

5. HOW MANY POINTS ARE ENOUGH?

For a *finite* data set one can argue that there is a distance (or radius) r below which there are no pairs of points (depopulation). At the other extreme, when the radius approaches the diameter of the cloud of points, the number of pairs no longer increases as the radius increases (saturation). The scaling region is always found between depopulation and saturation.

Because of Eq. (8.3), for a finite data set the population of pairs of points on smaller scales is smaller than the population of pairs on longer scales. This leads to poor statistics at small r. Because of the poor statistics, the function $C(r)$ may be distorted at small r appearing as in Fig. 80 rather than Fig. 79b. Nevertheless, provided that there are enough points, the scaling region over larger r's should remain unchanged.

Furthermore, if for a fixed n the number of points in the set becomes smaller, the population of pairs over the scales for which Eq. (8.3) hold begins to be depleted. As we continue to decrease the number of points (always for our fixed n), we observe

a. more and more depletion at smaller scales (since fewer and fewer points are found), and
b. large fluctuations of $N(r, n)$ due to small populations at larger scales.

The net result is that the scaling region may be completely masked. Any straight-line fitting at this point will give a false correlation dimension for that n. Thus, an accurate estimation of the slope d_2 on a $\log N(r, n)$ versus a $\log r$ plot requires a *minimum* number of points.

The scaling region can be similarly masked by increasing the embedding dimension while holding the number of points fixed. Then as n increases, the population density of the points must increase, implying that points tend to grow further apart and that an increasing number of balls of radius r centered on certain points will contain empty regions of the phase space. Thus, at some embedding dimension the scaling region is not clearly defined because it is "lost" between depopulation and saturation.

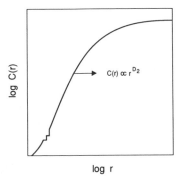

FIGURE 80.   For a finite data set the population of pairs of points over very small scales is smaller than the population of pairs over longer scales. This leads to poor statistics, which cause uncertainties in the function $C(r)$ over very small r's.

The embedding dimension for which the scaling region cannot be accurately defined is called the critical embedding dimension n_c. The foregoing points were emphasized by Essex et al.[49] and Tsonis and Elsner.[213]

We now wish to demonstrate these points by starting with a time series of 500 uniformly distributed random numbers in the interval [0, 1]. We know that for any embedding dimension n, $d_2 = n$ (as long as we use the necessary number of points for each n). We started with embedding dimension $n = 2$ and found $\log N(r, n)$ as a function of $\log r$; then we calculated

$$\text{Slope} = \frac{\Delta \log N(r, n)}{\Delta \log r}$$

as a function of $\log r$.

If there exists a clearly defined scaling region in the $\log N(r, n)$ versus $\log r$ plots, then we should be able, on a slope versus $\log r$ plot, to observe a plateau. This plateau will provide an estimate of the exponent d_2 for a given n. Figures 81a,b,c show slope versus $\log r$ for $n = 2$, $n = 6$, and $n = 15$, respectively. For $n = 2$ we observe that the slope has a nearly constant value of 2 for a wide range of scales. When the scales become too large, we see saturation indicated by a gradual decrease of slope. Depopulation is not visible (at least within the scale range of the figure). Therefore, we may conclude that 500 points are adequate in defining the scaling region when $n = 2$.

For $n = 6$ we observe a very different picture. We see depopulation, manifesting itself as many zero-slope values over smaller scales, large fluctuations over larger scales, and saturation over very large scales. A scaling region can be suggested, but it is not as clearly defined or as wide as for $n = 2$. Nevertheless, this small plateau is found at about slope 5.0, which is less than the true value of 6.0. Thus, an attempt to define a scaling region results at best in an underestimation of the true value of d_2. Similar comments can be made for $n = 15$; here the figure shows that there is virtually no way one can define a scaling region.

Figure 82 is similar to Fig. 81 but for $n = 4$. Similar comments to those made in Fig. 81 for $n \geq 6$ can be made for Fig. 82. A scaling region may be identified, but it will produce a value of d_2 close to 3.5 (which is less than the true value of 4.0). However, in Fig. 82b (with 5000 points), a well-defined scaling region exists with a value of $d_2 = 4.0$. The true scaling region here is at smaller scales compared to the scaling region identified when 500 points are used (where the true scaling region is masked by large fluctuations).

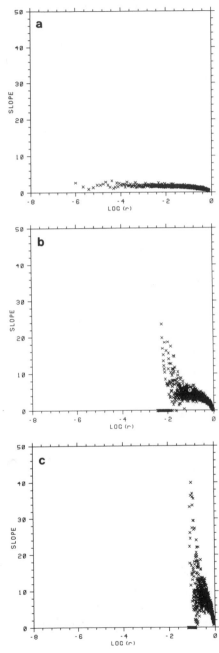

FIGURE 81. Slope $[\Delta \log N(r, n)/\Delta \log r]$ as a function of $\log r$ for embedding dimension (a) 2, (b) 6, and (c) 15. Based on a record of 500 uniformly distributed random numbers in the interval [0, 1]. (Tsonis *et al.*[215])

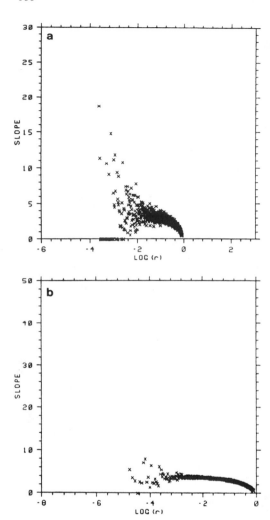

FIGURE 82.   (a) As in Fig. 81 for the embedding dimension 4. (b) As in Fig. 2a but for 5000 random numbers. (Tsonis *et al.*[215]).

The figures clearly demonstrate what was discussed previously. We should be very careful not to exceed the critical embedding dimension, which is a function of the data size. This unavoidably brings us to the following question: What is the necessary number of points N for a given embedding dimension? This problem can be approached by assuming that in all embedding dimensions (n) less than the dimension of the object in question, the object is space filling, as uniformly distributed random numbers in [0, 1]. The painful exercise of determining the minimum number of points as a function of the embedding dimension n was first tackled by Smith,[198] who concluded that this number is equal to 42^n. Thus, for an embedding dimension $n = 4$ if N is not equal to at least 3,111,696 no accurate estimate of d_2 can be obtained. Such a restrictive figure effectively "killed" all hopes for estimating the dimension of low-dimensional attractors irrespective of the availability of data, since even supercomputers could not handle such vast samples.

However, from our discussion of Figs. 81 and 82, accurate estimates for $n = 4$ can be obtained with as few as 5000 points. Why this great discrepancy? A possible explanation is that the 42^n conclusion is in error. In fact, it has been recently demonstrated[151] that Smith's procedure for obtaining the 42^n estimate is flawed and that the data requirements are not nearly as extreme. According to Nerenberg and Essex,[151] the minimum number of points N_{min} required to produce no more than an error A (typically $A = 0.05n$) is

$$N_{min} = \frac{\sqrt{2}[\Gamma(n/2 + 1)]^{1/2}}{(A \ln k)^{(n+2)/2}} \left[\frac{2(k - 1)\Gamma((n + 4)/2)}{[\Gamma(1/2)]^2\Gamma((n + 3)/2)}\right]^{n/2} \frac{n + 2}{2} \quad (8.4)$$

where n is the embedding dimension and $\Gamma(x)$ is the gamma function. The parameter k indicates how wide the scaling interval is. Recall that d_2 is estimated by the relation $d_2 = [\ln C(r') - \ln C(r)]/[\ln(r') - \ln(r)]$. By definition $k = r'/r$. For $A = 0.05n$ (in other words, for 95% confidence) and for $k = 4$ Eq. (8.4) can be approximated for $n < 20$ by

$$N_{min} \propto 10^{2+0.4n} \quad (8.5)$$

Thus, for $n = 4$, $N_{min} \sim 10^{3.6} \sim 4000$ points, as suggested by the direct analysis in Fig. 82. If fewer than N_{min} points are used, an error in estimating d_2 (underestimation) is expected. This error is given by inverting Eq. (8.4) to obtain

$$|\Delta d_2| = |d_2^{estimated} - d_2^{actual}|$$

$$= \left[\frac{(k - 1)^{n/2}(n + 2)^{n/2+1}\Gamma(n/2 + 1)^{n/2+1/2}}{\sqrt{2}N[\ln k]^{n/2+1}[\Gamma(1/n)]^n[\Gamma(n/2 + 3/2)]^{n/2}}\right]^{2/(n+2)} \quad (8.6)$$

As demonstrated in Figs. 81 and 82, Δd_2 is surely negative, since $d_2 = 3.5$ for $N = 500$ and $n = 4$, and $d_2 = 5.0$ for $N = 500$ and $n = 6$. In agreement with these results, we find from Eq. (8.6) that the expected error in estimating d_2 is approximately 0.45 and 0.9, respectively.

Keeping this in mind, assume that we have an object of dimension 4.3. We can take cognizance of the fact that in all embedding dimensions $n < 4.3$ the object is the space filling. Thus, for $n < 5$, $d_2 = n$ while for $n \geq 5$, $d_2 = 4.3$. Thus, the first deviation from the diagonal should provide an estimation for the correlation dimension (Fig. 83).

This, however, is not what is usually observed when data from measurements or from known dynamical systems are analyzed. What is commonly reported are graphs such as Fig. 84. The graph shows (dots) the correlation dimension d_2 as a function of the embedding dimension n for

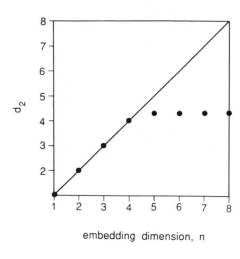

embedding dimension, n

FIGURE 83. Expected graph of the correlation dimension d_2 versus the embedding dimension n for an attractor of a dimension of 4.3 (see text for details).

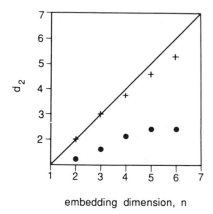

embedding dimension, n

FIGURE 84. Correlation dimension d_2 versus embedding dimension n (a) for data representing pulse of storm rainfall (points), and (b) a random sample of uniformly distributed numbers in the interval 0–1 (crosses). (Tsonis et al.[215]).

data representing pulses of storm rainfall (a time series of the time Δt between successive rain gauge signals, each corresponding to the collection of 0.01 mm of rain). Here the calculation of d_2 at higher and higher embedding dimensions is not achieved by successively shifting one time series by a delay parameter, but by introducing a new event (i.e., an independent time series of the same meteorological convective character) every time we go to a higher embedding dimension. Thus, the generated points in an n-dimensional space are independent, and the need of defining a suitable τ is overcome. The crosses are explained later.

What we observe in Fig. 84 is a gradual approach to a saturation value that is achieved at an embedding dimension $n_s = 5.0$ and that is estimated at 2.2. Surely, the first deviation from the diagonal does not correspond to the dimension of the underlying attractor (a feature often displayed in works employing the algorithm). In this example every rain event had 2200 points. Having such an example, can we argue that the estimated dimension ($D_2 \sim 2.2$) is correct? Let us proceed one step at a time. At embedding dimension 2 we already have a deviation from the diagonal, and 2200 points are plenty even by the most restrictive requirements. Why then does this deviation ($d_2 \sim 1.17$) not coincide with the saturation value 2.2? According to the previous arguments, if the dimension of the attractor is 2.2 then d_2 at $n = 2$ should be equal to 2, not 1.17. Since the number of points is much higher than required, there can be three explanations for what we observe: (1) Either the slope is systematically underestimated because the correlation function is convex; (2) we are looking at some unknown "artifact" of the algorithm. This artifact might arise in specific examples exhibiting properties that have not been investigated in detail with this algorithm (for example, in cases involving complicated, highly nonuniform probability functions); or (3) Whitney's embedding theorem applies. Accordingly, only a 6D phase space is sufficiently large to embed a 2.2-dimensional attractor. Thus, the value of d_2 may not be equal to 2.2 for $3 \leq n \leq 5$ and it may not be equal to n for $n \leq 2$. It is, therefore, quite possible that the first deviation from the diagonal would not give us the correlation dimension for which we may need to go to higher embedding dimensions.

This unavoidably brings us to the next very important point. *Because of the underlying assumptions, all theoretical calculations and derivations of the necessary number of points as a function of embedding dimension n presented thus far are valid only as long as $n < D_2$.* It is quite possible that estimates of the number of points depend on the type of attractor (nonuniform, fractal, etc.), an issue not yet addressed in these calculations. Recently, Lorenz[138] considered a chaotic dynamical system of dimension about 17 and applied the Grassberger-Procaccia algorithm to 4000 values of a selected variable from that system. He found that if $N < N_{min}$, then usually the dimension is underestimated. However, (1) different variables can yield different estimates, and (2) a suitably selected variable can sometimes yield a fairly good estimate. Such a variable is strongly coupled to the rest of the variables of the system. If this result holds for all dynamical systems, then it is more important to try to identify those strongly coupled variables than to worry about data requirements. Checking the results of the algorithm for different sample sizes may thus prove extremely useful.

Over and above those important issues, it is not yet known, especially for fractal sets, whether the need for data increases at the same rate with embedding dimension when $n > D_2$. Experimentation with known dynamical systems (Tsonis et al.[215]) indicates that even though the need for data may increase it may not be as severe as predicted by the formulas presented here. For example, Fig. 85 shows slope versus $\log r$ plots of a sample of size 2000 for an observable from the Hénon map for embedding dimensions 2, 4, 6, 8. We observe that for $n = 2$, $d_2 = 1.25$ (which is the Hausdorff-Besicovitch dimension of the Hénon map). Subsequently, we observe that up to $n = 8$, fairly accurate estimates of the correlation dimension can be obtained. In fact, we do not observe too much fluctuation or an underestimation of d_2 with increasing embedding dimension. Of course, eventually (i.e., for $n \gg D_2$) fluctuations may mask any scaling region. These results provide some evidence that for fractal sets, the need for data may increase at a much slower rate for embedding dimensions higher than the correlation dimension of the attractor.

If this is so, then in cases where saturation is observed for $n_s > D_2$, we may need $N \sim f(D_2)$ points rather than $N \sim f(n_s)$ points (again provided that n_s is not much higher than D_2). However, this conclusion may not be valid for every dynamical system or data set.

As a comparison, it has become a common task to apply the Grassberger-Procaccia technique to a random sample of size equal to the size of the observable in question. The random sample preferably has statistical properties similar to the observable (such as probability distribution, power spectra, etc.). We know that a random sequence results in "clouds" of points that do not exhibit any spatial correlation. In such cases, for any embedding dimension one should expect $d_2 = n$ (for a sufficiently large data set). Thus, in a d_2 versus n graph the correlation dimension estimates from the Grassberger-Procaccia algorithm would fall close to the diagonal (crosses in Fig. 84). Such a procedure provides some confidence that the saturation is not a result of underestimation due to an insufficient number

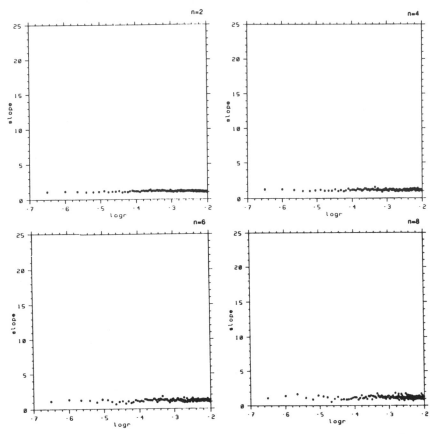

FIGURE 85.   Slope versus log$r$ for an observable from the Hénon map ($x_{t+1} = 1 - ax_t^2 + y_t$, $y_{t+1} = bx_t$, $a = 1.4$, $b = 0.3$). (Tsonis et $al.$[215])

of points (especially when saturation is observed for the observable but not for the random sequence).

Ruelle[185] has derived a simple formula for data requirements. For an embedding dimension n the slope in a log$N(r, n)$ versus log$r$ is

$$\frac{\log_{10}(N'', n) - \log_{10}N(r', n)}{\log_{10}r'' - \log_{10}r'}$$

where $r_{min} \leq r' < r'' \leq r_{max}$. Considering that $N(r', n) \geq 1$ and $N(r'', n) \leq 0.5N(N - 1) < N^2$ (note that N is the available number of points and

$N(r, n)$ is the number of pairs with distance less than or equal to r), it follows that

$$\log_{10}N(r'', n) - \log_{10}N(r', n) \le \log_{10}N^2$$

If this slope is measured over at least one decade, then $r'' \ge 10r'$ (where $r''/r' = k$), and thus

$$\log_{10}r'' - \log_{10}r' \ge \log_{10}10$$

or

$$\text{Slope} \le 2 \log_{10}N$$

In general, one can replace 10 by some number a and obtain

$$\text{Slope} \le 2 \log_a N \qquad (8.7)$$

According to Ruelle, this general formula provides an upper bound for D_2, and if $D_2 \le 2 \log_a N$ the estimate can be trusted. As pointed out by Essex and Nerenberg,[51] however, this bound is a bound of D_2 only if $k > a$, for an arbitrary value of a. When $k < a$, the right-hand side of Eq. (8.7) is *not* a bound. Since the value of a cannot be justified *a priori*, Essex and Nerenberg[51] conclude that Eq. (8.7) is of no practical use.

We make a final comment on the effect of noise in the data. Noise (or measurement error) acts as an infinite-dimensional system. It thus diffuses the fractal structure of the attractor. In general, when there is an external noise of a given magnitude, a plot of $\log C(r)$ versus $\log r$ has two

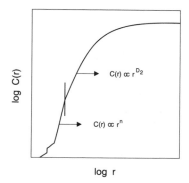

FIGURE 86.   The effect of noise in the data is seen in this figure. For scales greater than the amplitude of the noise the fractal structure is not affected significantly, and we observe the true scaling $C(r) \propto r^{D_2}$. For scales smaller than the amplitude of the noise the fractal structure is distorted and the slope approaches the embedding dimension. We now observe the scaling $C(r) \propto r^n$.

scaling regions (see Fig. 86). For length scales above those on which the noise significantly distorts the fractal structure, $C(r)$ scales like r^{D_2}. On scales below those affected by the noise, $C(r)$ scales like r^n. Consequently, the analysis of noisy data *may* simultaneously provide an estimate of the underlying dimension and of the size of the random component.

However, if the noise is considerable, the attractor will occupy the total available space and the scaling region will disappear. Subsequently, estimated dimensions in the presence of noise can range from the actual dimensions to infinity. Thus, noise may be quite a problem when we are dealing with observables from an uncontrolled experiment (atmosphere, brain, etc.). Reducing the noise level in the data has lately become a major task in the area of dynamical systems research (see Chapter 11).

6. DISTINGUISHING CHAOTIC SIGNALS FROM NONCHAOTIC OR FROM RANDOM FRACTAL SEQUENCES

Assume we have an arbitrary large data set free of noise. Further assume that we estimate a correlation dimension for this set that is finite and fractal. What does this finite dimension really mean? Can we conclude that there is a chaotic underlying attractor?

The importance of this question is that, given a strange attractor, a finite correlation dimension follows, whereas the opposite statement may not be true. As many will argue, there is no theorem that says given a finite correlation dimension an associated system has a strange attractor! To complicate things, counterexamples of certain Hamiltonian and/or random systems yielding finite correlation dimensions have been presented.

Benettin et al.,[16] Osborne and Caponio,[155] Chernikov et al.,[33] Bishop and Lomdahl,[22] Bishop et al.,[21] and others have presented examples of Hamiltonian systems having a small finite correlation dimension. As we know, Hamiltonian systems cannot have an attractor.

In addition, Osborne et al.[156] and Osborne and Provenzale[157] have argued that a certain class of *random* sequences exhibit a finite correlation dimension. This class includes self-affine sequences that exhibit a power-law spectra of the form $P(f_k) = Cf_k^{-a}$ for $1 < a < 3$, and are commonly referred to as fractional Brownian motions (fBm's) or colored noise. In this case the *trail* of n independent realizations (each one representing one phase-space dimension—see, for example, Fig. 19) is self-similar with a theoretically predicted fractal dimension of $2/(a - 1)$. When the Grassberger-Procaccia algorithm is applied to trails or to trajectories reconstructed from a single sequence via the method of delays, a finite correlation dimension (close to the fractal dimension of the trail) is obtained. Thus, they suggested that a finite value for the correlation dimension cannot indicate a dynamical system with a finite number of degrees of freedom. It can only indicate a lower bound for the actual number of degrees of freedom, which might be infinite! For the first time it was shown that the algorithm could not differentiate between chaos and colored noise.

As it turns out, however, the observation of Osborne and Provenzale may have no relevance to the practical estimation of dimensions from time series. We must note that the concept of fractal dimension can be applied to time series in two distinct ways. The first is to indicate the *number of degrees of freedom* in the underlying dynamical system. The second is to quantify the *self-similarity* of the trajectory in phase space. The Grassberger-Procaccia algorithm yields the first one. It does not really provide an estimate of the self-similarity of the trajectory. For example, applying the algorithm to an observable from the Lorenz system results in a dimension of about 2.07. This value has nothing to do with the self-similarity of the Lorenz trajectory (which by the way is not self-similar). Theiler[206] attacked this issue in an analytical way and proved that Osborne and Provenzale's anomalous scaling would not have been observed had they used the required number of points (which in the case of fBm's is very large due to their very long correlations) or if they had evaluated the correlation integral for smaller r's. Instead, the scaling $C(r, n) \propto r^n$ would have been observed. His results are summarized in Fig. 87, which shows the correlation integral for embedding dimension $n = 1, 2, \ldots, 15$ using $N = 16,384$ points in a time series of $1/f^2$ noise with high-frequency cutoff $f_1 = 0.05$ and low-frequency cutoff $f_0 = 0.00006$. One observes $C \propto r^n$ for small r, $C \propto r$ for a very small range of intermediate values of r, and the anomalous scaling $C \propto r^{2/(a-1)}$ for very large r.

These developments raise several questions as far as the application of the algorithm is concerned. Can the algorithm really be fooled, or is it simply not properly applied? *Most importantly, what are Hamiltonian systems and/or fBm's good for, in the interpretation of data such as those sampled from physical systems?* Hamiltonian systems in nature are at best rare (Chapter 3), and fBm's are nonstationary processes (Chapter 4). For fBm's the autocorrelation function $C(\xi) \propto \xi^{1-a}$ for $N \to \infty$. Therefore, $C(\xi)$ never reaches zero. Thus, no pairs are really independent. In addition, for finite N the decorrelation time is a function of N. Because of their properties, Mandelbrot[141] remarked that fBm's are not effective candidates

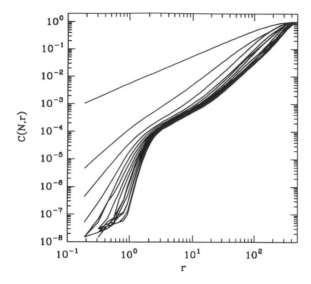

FIGURE 87. Correlation integral calculated from 16,386 points in a time series of $1/f^2$ noise with high-frequency cutoff $f_1 = 0.05$ and low-frequency cutoff $f_0 = 0.00006$. Results for embedding dimensions $n = 1, 2, \ldots, 15$ are shown. For small r the scaling obeys the equation $C \propto r^n$. For a small range of intermediate values of r the scaling becomes $C \propto r$. For very large r a scaling $C \propto r^{2/(a-1)}$ is observed. (Figure courtesy of Dr. James Theiler.)

for modeling natural processes. If any natural process were an fBm, it would have by now grown enough to destroy nature. Therefore, by definition natural processes are not fBm's and cannot exhibit $1/f^a$ spectra. The argument that some data might, during a finite time, mimic an fBm can be dealt with by simply testing the data for stationarity or by looking at the autocorrelation function for various lengths of the record or by making sure that the data set is longer than the length of a dominant period or by nonlinear prediction (see Chapter 10).

These unsettled issues may make dimension estimates difficult to accept. That is why the estimation procedure must address all points raised in the previous sections. A recommended procedure is outlined in Table II. Furthermore, additional evidence, such as testing for nonlinearity, dimension estimates using other algorithms, and Lyapunov exponents, should be provided.

TABLE II
Recommended Procedure To Estimate the Correlation Dimension

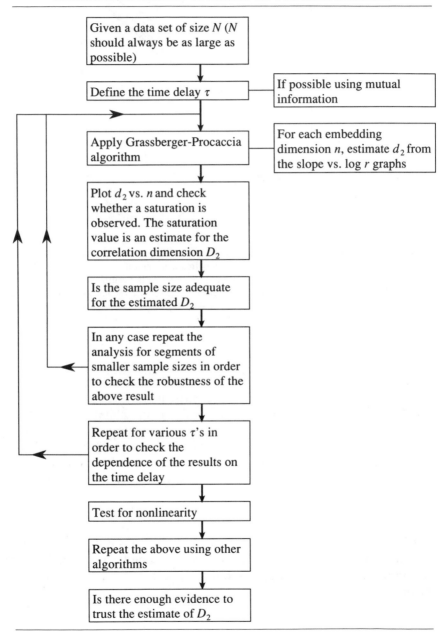

7. TESTING FOR NONLINEARITY

Assuming that strong statements can be made that we are dealing with samples from dissipative systems, we can argue that if the necessary number of points has been used then dimension estimates can be trusted. When, however, we cannot have the necessary number of points, we may wish to test the significance of our results. In fact, testing the results may be a good idea even if we have enough points.

The best way to test the significance of our dimension estimate is to produce a figure such as Fig. 84 for our data and for i.i.d. (independent identically distributed) processes or to use a test such as the BDS (Brock, Dechert, and Scheinkman) test. For this test (see Brock[24]) the BDS statistic is defined by

$$\text{BDS}(n, N, r) = \sqrt{N}[C(n, N, r) - C(1, N, r)^n]$$

where N is the number of points and n is the embedding dimension. If the time series is an i.i.d. process, then the BDS statistic converges (for large N) to a normal distribution with zero mean and a variance $V(n, r)$, which is independent of N. If the data are correlated (linearly or nonlinearly), then $C(n, N, r) > C(1, N, r)^n$ and the statistic diverges as $\sqrt{N}$.

Another approach (see Casdagli[29]) is the method of surrogate data. Given a time series, create i surrogate time series that are random but contain the same linear correlations as the original raw data. A practical way to do this is to take a Fourier transform of the raw data, randomize the phases, and then take the inverse Fourier transform. Proceed to compute the statistic x_{si} (dimension or Lyapunov exponent) for each surrogate time series and the same statistic x_r for the raw data. Let μ_s and σ_s be the mean and standard deviation of the statistic x_s. If the difference $|\mu_s - x_r| \gg \sigma_s$, then one can be confident that there is significant nonlinear structure in the raw data that is not apparent in the linear stochastic surrogate data.

A similar test was recommended by Osborne et al.[156] to test whether the raw data are fBm's. The difference was that they first approximated the raw Fourier transform with a power-law spectrum, and then generated a time series by inverting the $1/f^a$ spectrum with the procedure outlined in Chapter 4. Such a test, however, assumes that power-law spectra adequately describe the spectra of natural processes, which, as discussed in the previous section, may not be a realistic assumption. In addition, as Theiler[206] notes, such a procedure, by construction, fails to reject the null hypothesis that the original time series is not an fBm.

8. OTHER APPROACHES TO ESTIMATE DIMENSIONS

While the Grassberger-Procaccia algorithm is the most popular, other effective approaches are available. Badii and Politi[13] developed a method for the investigation of fractal attractors based on the properties of nearest neighbors on the attractor. The structure of an attractor is characterized by the way points on the attractor are distributed. For a given attractor the distribution of neighbors around a point on it is thus dictated by the underlying dynamics. According to Badii and Politi,[13] the average distance between a point and its nearest neighbor, $\langle \delta \rangle$, scales as a function of the number of points included in the averaging according to

$$\langle \delta \rangle \sim n^{-1/D}$$

where D is a suitable dimension. This relation can be extended to the generic moment of order γ:

$$\langle \delta^{\gamma} \rangle = K n^{-\gamma/D(\gamma)}$$

where $D(\gamma)$ is a γ-dependent definition of dimension called the dimension function:

$$D(\gamma) = -\lim_{n \to \infty} \frac{\log n}{\log \langle \delta^{\gamma} \rangle}$$

The dimension function is related to the various dimensions via the relation

$$D(\gamma = (1 - q)D_q) = D_q$$

Thus, for $q = 0$

$$D(D_0) = D_0$$

for $q = 1$

$$D(0) = D_1$$

for $q = 2$

$$D(-D_2) = D_2$$

and so on. Here D_0, D_1, and D_2 are the capacity dimension, the information dimension, and the correlation dimension, respectively.

The algorithm to estimate all these dimensions is straightforward. First, for a given γ the curves $\log\langle \delta^\gamma \rangle$ versus $\log n$ are produced. These curves yield an estimate for $D(\gamma)$. For many values of γ a graph $D(\gamma)$ versus γ can be produced that will provide D_0, D_1, and D_2. Badii and Politi note that these relations hold not just for the nearest neighbor but for the second nearest, the third nearest, etc. They recommend using the third nearest neighbor, because they find the sensitivity to statistical fluctuations to be much weaker for the second and third nearest neighbors compared to the first one.

Another interesting approach is based on unstable periodic orbits. Recall from Chapter 5 that the unstable periodic orbits make up the "skeleton" of the attractor. The complete set of these orbits is thus related to the structure of the attractor and, consequently, to its invariant measures, such as dimensions and Lyapunov exponents.[95]

In practice, periodic orbits are found by preassigning a small spatial distance r and by finding how many time steps n it takes for the time series to come within a distance r from a given point. The number n indicates a cycle of period n. Given a large amount of points N, all the period n orbits can be cataloged.[10] As Auerbach et al.[10] and Cvitanovic et al.[39] demonstrate, the knowledge of the number of periodic orbits of order n can be used to infer the topological entropy of a dynamical system as well as its dimensions and Lyapunov exponents. Recently, Lathrop and Kostelich[122] have described a general procedure to locate periodic saddle orbits from experimental data and to calculate the Lyapunov exponents.

9. ESTIMATING THE LYAPUNOV EXPONENTS FROM TIME SERIES

In Chapter 5 we discussed how to obtain the largest Lyapunov exponent of a known dynamical system (Benettin et al.[17]) or its complete Lyapunov exponent spectrum (Jacobian method). Both techniques have now been developed into algorithms for estimating Lyapunov exponents from observables.

Wolf et al.[225] presented an approach by which the largest exponent, λ_1, is estimated once the attractor has been reconstructed. In theory, λ_1 is estimated by monitoring the long-term evolution of a single pair of nearby orbits. The reconstructed attractor, however, contains just one trajectory.

The reconstruction can, nevertheless, provide points that may be considered to lie on different trajectories if we choose two points whose temporal separation in the original time series is at least one mean orbital period (orbital periods may be defined by the Fourier spectra). As long as their spatial separation in the reconstructed attractor is small, those two points may be considered to define the early state of the first principal axis. We then monitor their separation, and when it becomes large we replace the nonfiducial point with a point closer to the fiducial point (in the same manner outlined in Fig. 43). We then monitor the separation of these two new points. Repeating this procedure many times leads to an estimate of λ_1. Recently, Frank[65] modified the Wolf et al.[225] algorithm for improved estimation of λ_1 in cases of noisy and small data sets.

According to Jacobian methods,[45,189] a neighborhood of m points within a small distance is considered around a reference point on the fiducial trajectory, and then a local linear map that maps the whole neighborhood into a neighborhood after some time T is derived. To obtain the mapping, the location of the neighbors at each time step is monitored. For sufficiently small neighborhoods and time intervals, the evolution of nearby states is approximated by equations such as $x'_{n+1} = Ax'_n$ (for maps) or $x'(t) = e^{tA}x'(0)$ (for flows). Therefore, information about local state-space expansion and contraction rates is contained within the *linearized* equations.

This provides the reasoning behind obtaining local linear maps. Take as an example the logistic map $x_{n+1} = 4x_n(1 - x_n)$. In linearized form we have $x'_{n+1} = ax'_n$ with $a = -2$ [see Eq. (6.8)]. Thus, small fluctuations amplify in time. From this equation one can now deduce the expansion rate, which is equal to $|2|$ [recall that the equation $x'_{n+1} = ax'_n$ is the same as Eq. (6.9)]. Therefore, the constant a is the slope whose absolute value is related to the spreading of nearby trajectories. As a consequence, the Lyapunov exponent $\lambda_1 = \log2$, in agreement with its theoretical value.[184]

When we try to obtain a mapping from a reconstructed attractor, we basically assume that the m points are small fluctuations from the reference point and that the evolution of each fluctuation obeys a linear law. For each point a value for a is obtained. For many points the idea is to find an optimal a that minimizes the linear fit error from all neighbors. Extending the arguments to n dimensions leaves us with the task of estimating an $n \times n$ matrix A whose eigenvalues provide the Lyapunov exponents. The problem, however, is how to do this when the mapping is not known, for example, when we are dealing with an observable from an unknown dynamical system.

Let us assume that we observe the time series $\{s(t)\}$: 1, 1, 2, 3, 1, 2,

2, 3, 4, 5, 3, 3, 2, 1. As Table III illustrates, we may reconstruct a 3D phase space whose coordinates are $x_1(t) = s(t)$, $x_2(t) = s(t + \tau)$, $x_3(t) = s(t + 2\tau)$ with $\tau = 1$. Such a reconstruction results in a sequence points $\{y(n)\}$. In our example, $y(1)$ has coordinates $[1, 1, 2]$, point $y(2)$ has coordinates $[1, 2, 3]$, etc. Note that the parameter n in $y(n)$ plays the role of time and determines the order with which the points appear on the attractor. We define the nearest neighbor to point $y(1)$ as $y^1(1)$ (thus, a superscript k indicates the kth closest neighbor). This is the point with coordinates $[1, 2, 2]$. We define the distance between $y(1)$ and $y^1(1)$ as $z^1(1)$. Let us assume that there is some underlying mapping whose action takes $y(1)$ and $y^1(1)$ and moves them to the next time step [i.e., to points $y(1 + 1)$ and $y(1, 1 + 1)$, respectively]. Note that $y(1, 1 + 1) \neq y^1(1 + 1)$. In other words, the closest neighbor to $y(1 + 1)$ is *not* necessarily point $y(1, 1 + 1)$. Let us denote the distance between $y(1 + 1)$ and $y(1, 1 + 1)$ by $z(1, 1 + 1)$. Then we have

$$z(1, 1 + 1) = F(y^1(1)) - F(y(1))$$
$$= F(y(1) + z^1(1)) - F(y(1)) \qquad (8.8)$$

We can generalize Eq. (8.8) for any point $y(n)$ and any of its $r = 1, \ldots, m$ closest neighbors:

$$z(r, n + 1) = F(y(n) + z^r(n)) - F(y(n)) \qquad (8.9)$$

TABLE III

This Hypothetical Example of a 3D Phase Space Reconstruction Is Used To Define Nearest Neighbor and the Distance between a Point and Its Nearest Neighbor

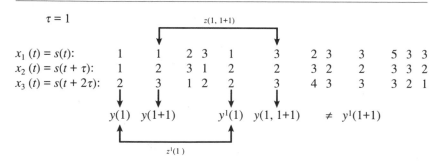

By expanding Eq. (8.9) in a Taylor's series about $z^r(n)$ and keeping only first-order (linear) terms, we arrive at

$$z(r, n + 1) = Az^r(n) \qquad (8.10)$$

where

$$A = \begin{pmatrix} \dfrac{\partial F_1}{\partial x_1} & \dfrac{\partial F_1}{\partial x_2} & \dfrac{\partial F_1}{\partial x_3} \\[2mm] \dfrac{\partial F_2}{\partial x_1} & \dfrac{\partial F_2}{\partial x_2} & \dfrac{\partial F_2}{\partial x_3} \\[2mm] \dfrac{\partial F_3}{\partial x_1} & \dfrac{\partial F_3}{\partial x_2} & \dfrac{\partial F_3}{\partial x_3} \end{pmatrix}$$

The reader may recognize that the matrix A is the 3×3 (the embedding dimension in our example is 3) Jacobian matrix (recall our discussion in Chapter 2).

Since $z(r, n + 1)$ and $z^r(n)$ are vectors, we can define the first component of $z(r, n + 1)$ as

$$z_1(r, n + 1) = \frac{\partial F_1}{\partial x_1} z_1^r(n) + \frac{\partial F_1}{\partial x_2} z_2^r(n) + \frac{\partial F_1}{\partial x_3} z_3^r(n)$$

or, considering all the neighbors, as

$$\begin{bmatrix} z_1(1, n + 1) \\ z_1(2, n + 1) \\ \vdots \\ z_1(m, n + 1) \end{bmatrix} = \begin{bmatrix} z_1^1(n) & z_2^1(n) & z_3^1(n) \\ z_1^2(n) & z_2^2(n) & z_3^2(n) \\ \vdots & \vdots & \vdots \\ z_1^m(n) & z_2^m(n) & z_3^m(n) \end{bmatrix} \begin{bmatrix} \dfrac{\partial F_1}{\partial x_1} \\[2mm] \dfrac{\partial F_1}{\partial x_2} \\[2mm] \dfrac{\partial F_1}{\partial x_3} \end{bmatrix} \qquad (8.11)$$

We abbreviate this as $A = BC$. The entries of matrix B are provided from the original time series. For our example, $z_1^1(1) = s(5) - s(1)$, $z_2^1(1) = s(6) - s(2)$, and $z_3^1(1) = s(7) - s(3)$. In general, $z^r(n) = \{s(n_r + (a - 1)\tau) - s(n + (a - 1)\tau)\}$, where n_r is the n value associated with the rth neighbor to $y(n)$ and $a = 1, 2, \ldots, (d_E - 1)$, where d_E is the embedding dimension. Similarly, the entries of matrix A are given by the expression $z_a(r, n + 1) = \{s(n_r + 1 + (a - 1)\tau) - s(n + 1 + (a - 1)\tau)]$. Thus, we can invert matrix B to obtain $\partial F_1/\partial x_1$, $\partial F_1/\partial x_2$, and $\partial F_1/\partial x_3$. We then repeat for the second and third components to obtain all the entries in matrix A. In practice, we have m equations in k unknowns with $m > k$. The problem is thus overdetermined, and usually the solution $C = AB^{-1}$ does not exist (B is not invertible). In such cases the solution is obtained by least-squares methods. One of those methods is outlined in Eckmann et al.[45] Note that, due to noise, we want the problem to be overdetermined. If it is not, we might not capture enough dynamics. In our example the embedding dimension was 3. All formulas can be generalized for any embedding dimension d_E. Then Eq. (8.11) becomes

$$
\begin{bmatrix}
z_1(1, n + 1) \\
z_1(2, n + 1) \\
\vdots \\
z_1(m, n + 1)
\end{bmatrix}
=
\begin{bmatrix}
z_1^1(n) & z_2^1(n) & \cdots & z_{d_E}^1(n) \\
z_1^2(n) & z_2^2(n) & \cdots & z_{d_E}^2(n) \\
\vdots & \vdots & & \vdots \\
z_1^m(n) & z_2^m(n) & & z_{d_E}^m(n)
\end{bmatrix}
\begin{bmatrix}
\dfrac{\partial F_1}{\partial x_1} \\[6pt]
\dfrac{\partial F_1}{\partial x_2} \\[6pt]
\vdots \\[6pt]
\dfrac{\partial F_1}{\partial x_{d_E}}
\end{bmatrix}
$$

Note that this procedure refers to just one point and one time step. In practice, for this point we estimate the $d_E \times d_E$ Jacobian for k steps along the orbit. Thus, we obtain $A(1), A(2), \ldots, A(k)$. Then according to the Oseledec multiplication ergodic theorem (see Abarbanel and Kennel[3]), the Lyapunov exponents (for the point and its neighbors) are the logarithms of the eigenvalues of the matrix

$$
\lim_{k \to 0} \{ [A^k]^{\mathrm{T}} [A^k] \}^{1/2k}
$$

where T signifies transpose and $A^k = A(1)A(2) \cdots A(k)$.

Finally, everything is repeated for many other points and their neighbors. The Lyapunov exponent spectrum for the dynamical system in question is produced by the average from all the points. However, as pointed out by Abarbanel et al.,[1,2] these approaches may not provide reliable estimates for all but the leading Lyapunov exponents. The difficulty in accurately determining the negative exponents arises because fractal attractors are often thin at many locations along directions of convergence of volumes in phase space (associated with negative exponents). Such thin regions make it difficult to define neighborhoods, not to mention that such fine structure may easily be distorted by noise in the data.

A way out of these problems is to consider a huge number of points. On the other hand, if we consider neighborhoods that are large compared with the thickness of the attractor yet small compared with the size of the attractor, then the points in these neighborhoods, in general, lie close to some curved subsurface within the local neighborhood. This curvature can cause severe problems when a linear mapping is employed. Following these points, Brown et al.[27] showed that nonlinear mappings are clearly advantageous in accurately estimating the Lyapunov exponent spectrum.

For any of the foregoing approaches numerous points are required for a good delineation of the underlying attractor. It has been initially suggested that between 10^d and 30^d points are needed. In view of the recent results concerning the number of points in dimension estimates, however, it is possible that these suggestions are on the high side. Unlike the procedure for estimating dimensions, here the calculations require just *one* embedding dimension d_E. Thus, it becomes critical to define a proper embedding dimension for the calculations. In general, the embedding dimension is higher than the "dynamical" embedding dimension d, which is the dimension of the dynamics underlying the measurements (this is the minimum embedding dimension the attractor can be embedded in). The reason for this practice is that if d_E is not large enough the embedding may cause the attractor to be folded up in such a way that it crosses itself in certain areas (see Fig. 88). If we then choose neighbors to a given point on the basis of their displacement in phase space, there will be no distinguishing the two parts of the reconstructed attractor in the intersection zone. In general, the self-intersections have dimension $2d_a - d_E$, where d_a is the fractal dimension of the attractor. If we have just one self-intersection (i.e., one point), then $2d_a - d_E = 0$ (the dimension of a point is zero). Thus, we can avoid self-intersections if we take

$$d_E > 2d_a$$

in agreement with the result by Takens.[203] Since $d_E > d$, we estimate more Lyapunov exponents than those that are really related to the dimension of the attractor. Those exponents are called spurious and do not provide any useful information about the dynamics of the underlying system (they are all negative). With respect to the choice of the delay in reconstructing

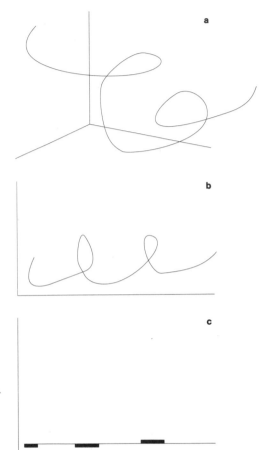

FIGURE 88.   (a) A line embedded in three dimensions. It loops around but does not cross itself. (b) A projection of the line in (a) to a two-dimensional space. The line now has self-intersections that are points (each of dimension zero). (c) Projection of the structure in (b) to the one-dimensional space. Now the line develops intersections that are line segments (each one having dimension 1). The situation in (c) is quite serious [more than in (b)] because the ability to distinguish true neighbors of a given point on the line in 3D has been lost for a large part of the line. Thus, a structure of dimension 1 (the line) requires embedding dimension $d_e = 3$ to avoid all self-intersections.

the attractor, the same comments made when discussing dimension estimations apply. Just recently, Abarbanel (personal communication) proposed an interesting new approach to define a proper embedding dimension from an observable. Starting with $d = 1$, for each point we find the nearest neighbor. From all available points we then find how many of those nearest neighbors remain nearest neighbors as we go to $d = 2$, $d = 3$, $d = 4$, and so on. A proper embedding dimension is defined as the dimension for which the percentage of false neighbors goes to zero.

The issue of the embedding dimension brings up another issue, noise in the data. The accuracy of the data also plays an important role in estimating the exponents. Obviously, noise introduces an uncertainty in the

TABLE IV
Recommended Procedure To Estimate the Lyapunov Exponents

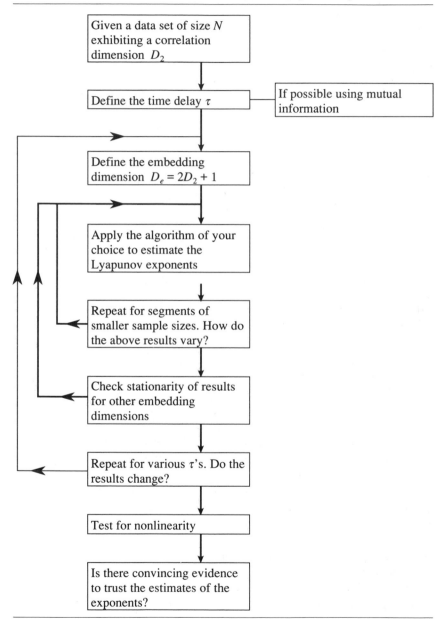

estimates. Brown *et al.* [27] concluded that the value of negative exponents in thin directions of the data start to become affected when the noise level is higher than 10% of the thickness of the data. The presence of noise also affects state-space reconstruction. [31]

Table IV presents a flowchart with the recommended procedure to estimate Lyapunov exponents. Recall from Chapter 5 that there exists a formula [Eq. (5.10)] that relates the Lyapunov exponents to the dimension of the attractor. Therefore, the algorithms used to calculate dimensions and Lyapunov exponents should, in principle, produce results that are compatible. In other words, the Lyapunov exponents estimates should yield a dimension close to that estimated from an algorithm. Otherwise an explanation is in order.

CHAPTER 9

EVIDENCE OF CHAOS IN "CONTROLLED" AND "UNCONTROLLED" EXPERIMENTS

Chapter 8 outlined the details behind the theory of phase-space reconstruction and the estimation of the various relevant exponents. The development of the theory and the various techniques were verified via well-defined low-dimensional mathematical dynamical systems. The next step was to see whether chaotic behavior is found in physical systems. The next best thing to mathematical dynamical systems would be a "controlled" physical system where only a small number of variables and parameters are important (controlled in the sense that the parameters that dictate the evolution of the system can be held fairly constant). To this end many intelligent experiments were set up, and the search for evidence of chaos in the laboratory was begun. Several classic controlled experiments will now be discussed in detail.

1. CHEMICAL EXPERIMENTS

The most celebrated oscillating chemical system involves the cerium-catalyzed bromination and oxidation of malonic acid by a sulfuric acid of bromate.[108,202] According to the experimental setup (see Fig. 89), malonic acid [$CH_2(COOH)_2$], potassium bromate ($KBrO_3$), cerium sulfate [$Ce_2(SO_4)_3$], and sulfuric acid (H_2SO_4) are continuously pumped at a rate r into a well-stirred reactor. The stirring makes the system homogeneous. The chemical mechanism involves more than 20 reactions and as many products. Due to the difficulties in recording more than two independent signals, the dynamics of this chemical experiment have been studied in

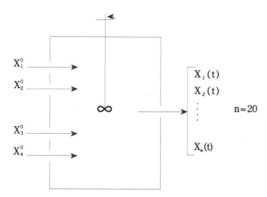

FIGURE 89. Experimental setup of the Belousov-Zhabotinsky reaction. Four chemicals are pumped into a well-stirred reactor. This reaction results in about 20 products.

terms of the observed concentration of one of the products (commonly bromide ion Br^{2+} or cerium ion Ce^{4+}). Such a reaction is referred to as the Belousov–Zhabotinsky reaction.

Figure 90 shows experimental results[218] as the residence time, t = reactor volume/flow rate, changed from $t = 0.49$ hr to $t = 1.03$ hr. The figure presents for each t the observable $V(t)$ (in this case bromide ion potential), the corresponding power spectrum, and a 2D phase portrait. The coordinates of the phase portrait are $V(t)$ and $V(t + \tau)$, where $\tau = 8.8$ sec. In Fig. 90a ($t = 0.49$ hr) the evolution is periodic of period 1. In the spectra the noise level is at or below -5. Above this level we observe peaks at multiples of one frequency (~ 0.0115). The phase portrait is a limit cycle. For $t = 0.90$ hr (Fig. 90b) the observable becomes "irregular." The spectra show many peaks superimposed as a continuous background that is significantly above the noise level indicated in Fig. 90a. The phase portrait shows a strange attractor. The evolution has become chaotic. As the residence time increases, the dynamics change again, and for $t = 1.03$ the evolution again becomes periodic but is now of period 2 (see Fig. 90c). According to the investigators, between any two periodic regimes there is always a chaotic regime.

In a similar experiment Simoyi et al.[197] concluded that period doubling can be observed in the Belousov-Zhabotinsky reaction, leading to a regime containing both chaotic and periodic states. In both the above-mentioned studies it is suggested that the chaotic attractor is essentially a

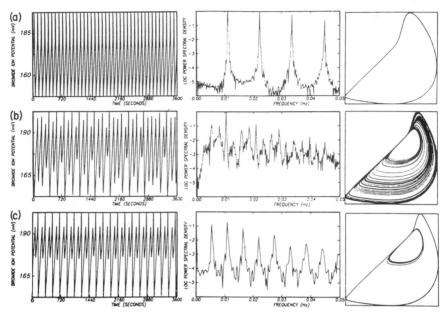

FIGURE 90.   Results for three residence times: (a) $t = 0.49$ hr, (b) $t = 0.90$ hr, (c) $t = 1.03$ hr. For each t the graphs show a time series of bromide ion potential $V(t)$, the corresponding power spectra, and 2D $[V_1 = V(t), V_2 = V(t + \tau)]$ portrait for $\tau = 8.8$. (Reproduced by permission from Turner *et al.*[218].)

two-dimensional (sheetlike) attractor, and thus a three-dimensional construction of the phase space is adequate. The same conclusion was reached earlier by Roux *et al.*,[179] reporting on a three-dimensional attractor in the Belousov-Zhabotinsky reaction.

Figure 91a shows a two-dimensional projection of the three-dimensional phase portrait taken from the experimental results of Turner *et al.*[218] and Roux *et al.*[181] [with the third axis $V(t + 17.6)$ normal to the page]. Figure 91b shows the Poincaré section formed by intersecting the three-dimensional phase space with a plane normal to the page and passing through the dashed line in Fig. 91a. We see that the section is quite simple and nearly one dimensional. We can thus construct a one-dimensional map by recording for successive intersection the values of $V(t)$ and plotting V_n versus V_{n+1}. The corresponding map is shown in Fig. 91c, where a solid line has been "eye" fitted. A very similar map was derived by Simoyi *et al.*[197] and by Hudson and Mankin[107] in their study of chaos in the Belousov-Zhabotinsky reaction.

For the chaotic regime, a positive Lyapunov exponent has been esti-

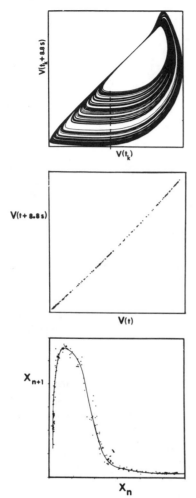

FIGURE 91. (a) Two-dimensional projection of a 3D reconstruction of the attractor in the Belousov-Zhabotinsky reaction. (b) By intersecting the 3D reconstructed attractor with a plane, we obtain the Poincaré section. (c) The section appears to be quite simple and nearly one dimensional. We can thus seek to construct a one-dimensional map by recording the values of $V(t)$ that correspond to successive intersections and then plot V_n vs. V_{n+1}. The solid line has been "eye" fitted. (Reproduced by permission from Dr. H. L. Swinney. Parts of this figure appeared in Roux et al.,[181] and Turner et al.[218].)

mated.[107,180,181] The value of $\lambda \sim 0.62$ is suggested. Intermittency has also been discovered in the reaction. Recall that intermittency is a route to chaos according to which periodic evolutions are interrupted by random bursts. Pomeau et al.[169a] demonstrated that for a residence time of about 76 min, stable oscillations in the Ce^{4+} concentration are interrupted from time to time and at random by large peaks (see Fig. 92). A review of other studies on several types of bifurcations and intermittency in the reaction is presented in Roux[177] and Roux et al.[178]

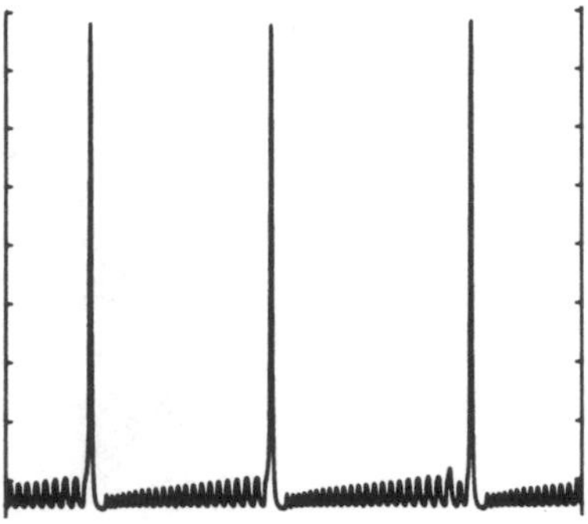

FIGURE 92. Intermittency observed in the Belousov-Zhabotinsky reaction. Periodic oscillations are interrupted by random bursts. (Reproduced by permission from Pomeau et al.[169a].)

Other important studies dealing with the reaction are Epstein,[48] Hudson et al.,[106] and Vidal et al.[220]

2. NONLINEAR ELECTRICAL CIRCUITS

Nonlinear electrical circuits make ideal candidates for examining different types of dynamical behavior. This is partly due to the fact that their characteristic frequencies are up to seven orders of magnitude higher than the frequency of a chemical oscillator. A simple nonlinear electrical circuit is shown in Fig. 93. The system is described via the equation $L\ddot{q} + R\dot{q} + V_C = V(t)$ or $L\ddot{q} + R\dot{q} + q/C = V_0 \sin 2\pi ft$, where q is the charge across the varactor, L is the inductance, R is the resistance, and V_C is the varactor voltage ($V_C = q/C$, where C is the capacitance). Thus, the system has 3 degrees of freedom: q, $\dot{q}$, and $2\pi ft = \omega t = \theta$.

By keeping f a constant 96.85 kHz, Testa et al.[204] studied the dynamical behavior of q or V_C for varying values of V_0. Their results can be summarized in Fig. 94, the bifurcation diagram. It shows the richness of

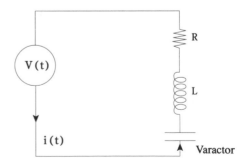

FIGURE 93. Experimental setup of a nonlinear electrical circuit with a varactor diode that conducts for a forward voltage and has a nonlinear capacitance for a reverse voltage.

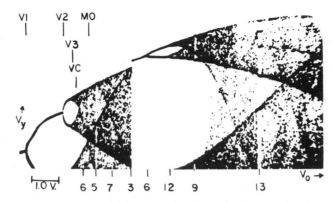

FIGURE 94. Bifurcation diagram corresponding to the above nonlinear electrical system. This diagram is produced by varying V_0 while keeping f equal to 96.85 kHz. The thresholds V_1, V_2, and V_3 correspond to evolutions of period 2, 4, and 8, respectively. Other thresholds such as V_c for chaos and windows of periods 6, 5, 7, 3, 6, 12, 9, and 13 are indicated. The veiled lines are peaks in the spectral density in the chaotic regime. (Reproduced by permission from Testa *et al.*[204].)

the bifurcation diagram of the logistic equation: period doubling leading to chaos, windows of periods 6, 5, 7, 3, 6, 12, 9, and 13, etc. In addition, Testa *et al.*[204] estimated that the scale of branch splitting $\alpha = 2.41$ and $\delta = 4.26$. These values are very close to theoretically predicted values of $\alpha = 2.5$ and $\delta = 4.67$ (see Chapter 6).

Period doubling and chaotic behavior were also observed by Linsay.[134] For a driven nonlinear semiconductor oscillator, which also showed a period-doubling pitchfork bifurcation route to chaos, the Pomeau-Manneville intermittency route was observed by Jeffries and Perez.[113] Periodic phases were interrupted by bursts of periodic behavior. Jeffries and Perez also estimated that the average duration of the chaotic interruptions scaled according to the equation $\Delta t \propto \epsilon^{-0.45}$, which is quite close to the theoretically predicted scaling $\Delta t \propto \epsilon^{-0.5}$ (see Chapter 6). Crisis has also been observed in nonlinear circuits.[112,174] Note that the first evidence for chaotic behavior in an electrical system was found by Gollub *et al.*,[81] in an experiment involving a system of two coupled oscillators.

3. COUETTE–TAYLOR SYSTEM

This system consists of two concentric cylinders that rotate independently with angular velocities V_1 and V_2 (see Fig. 95). A fluid is contained

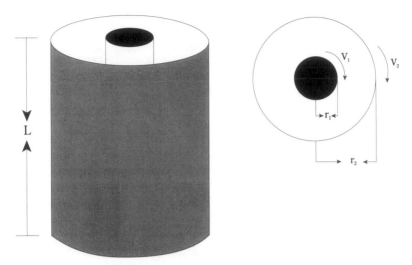

FIGURE 95.   Experimental setup of the Couette–Taylor system. The system consist of two concentric cylinders that rotate independently with angular velocities V_1 and V_2. The space between the cylinders is filled with a fluid.

between the two cylinders. Under such an arrangement the two Reynolds numbers are $R_1 = (r_2 - r_1)r_1 V_1/\nu$ and $R_2 = (r_2 - r_1)r_2 V_2/\nu$, where r_1 and r_2 are the radii of the inner and outer cylinders, respectively, and ν is the kinematic viscosity.

The Couette–Taylor system provides an excellent example of the periodic-quasiperiodic-chaotic route to chaos discussed in Chapter 6. For $V_2 = 0$ (i.e., the outer cylinder is not rotating), and for $R_1/R_c = 9.6$ (R_c is a constant) the dynamic behavior of the system is periodic, as the spectra in Fig. 96a show. These spectra correspond to a sequence of the local radial velocity of the fluid measured at a point. What we observe is a sharp peak superimposed on a continuous background that is due to instrumental noise. As the ratio R_1/R_c increases, for $R_1/R_c = 11.0$ we observe two fundamental frequencies in the spectra (Fig. 96b). A transition from periodic to quasiperiodic behavior has occurred. Further increase of R_1/R_c results in spectra that contain broadband noise well above the instrumental noise level in addition to the two sharp peaks. Now we have basically an aperiodic signal. The torus has given way to a chaotic attractor.

The periodic-quasiperiodic-chaotic transition is visualized in Fig. 97, which shows the observed flow progressively becoming more and more complex as the angular velocity R_1 is increased. The foregoing experiment

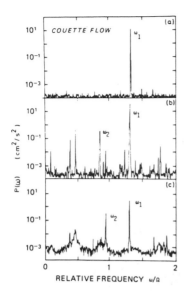

FIGURE 96. Spectra corresponding to an observable from the Couette–Taylor system for $R_2 = 0$ and varying R_1. The observable is the local radial velocity of the fluid measured at a point. As the Reynolds number R_1 increases, a transition from periodic (one fundamental frequency) to quasiperiodic (two fundamental frequencies to chaotic (broadband spectra) is observed. (Reproduced by permission from Swinney and Gollub[201].)

was designed and carried out by Gollub and Swinney[82] (see also Fenstermacher et al.[60]). The results challenged the Landau scenario for the onset of turbulence according to which an ever higher number of independent fluid oscillations should be excited as R_1 is increased. Once the second fundamental frequency is observed, further increase in R_1 would excite another independent frequency and then another, and so on. This is not what was observed. At some critical value R_c, a continuous range of frequencies appears. This observation is consistent with the presence of a chaotic attractor. This scenario for the onset of turbulence had been suggested by Ruelle and Takens,[186] who gave mathematical arguments suggesting that the attractor associated with the Landau scenario would not be a high-dimensional torus but a chaotic attractor, as originally postulated by Lorenz in 1963. For more information on the Couette-Taylor system, see Benjamin and Mullin,[18] Donnelly et al.,[43] Lorenzen et al.,[139] King and Swinney,[118] Shaw et al.,[195] and Swinney and Gollub.[201]

4. RAYLEIGH–BÉNARD CONVECTION

In a Rayleigh–Bénard system (see Fig. 98) a fluid is contained between parallel plates heated from below. The fluid at the bottom is heated, rises, and creates convection. The behavior in such a system is usually studied

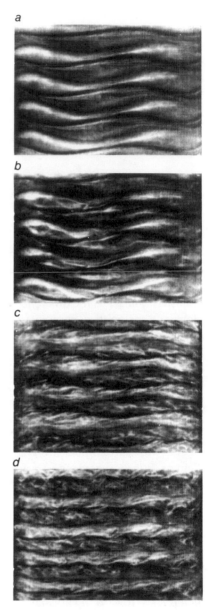

FIGURE 97. Shown here are successive pictures of the fluid in a Couette–Taylor system. The inner cylinder is rotated. The outer cylinder is not. As the angular velocity is increased, the flow becomes progressively more complex. (Reproduced by permission from Dr. H. L. Swinney. This figure appears in *Scientific American* **254,** 1986.)

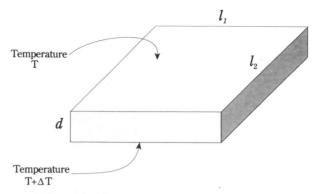

FIGURE 98.   Experimental setup of the Rayleigh–Bénard system. A fluid is contained between parallel plates. The bottom plate is heated, causing the fluid to rise and create convection.

as function of the Rayleigh number $R = (g\alpha d^3/k\nu)\,\Delta T$, where g is the gravitational acceleration, α is the thermal expansion coefficient, d is the separation between the plates, k is the thermal diffusivity, ν is the kinematic viscosity, and ΔT is the difference in temperature between the two parallel plates. Other control parameters are the Prandtl number $P = \nu/k$, the aspect ratios $\Gamma_1 = l_1/d$ and $\Gamma_2 = l_2$, and the boundary conditions at the side walls.

Figure 99 is similar to Fig. 96, but for the velocity at the center of the box. Again we observed that as the Rayleigh number increases, the periodic to quasiperiodic to chaotic transition takes place.[201] The period-doubling route to chaos, frequency locking, and intermittency has also been observed in the Rayleigh–Bénard system.[19,44,74,79,127,130] For example, Fig. 100 shows plots of vertical temperature gradient versus horizontal temperature gradient as R is increased. Period doubling and chaos are observed. The Feigenbaum number δ has been estimated in some of these studies. In one[127] δ is estimated to be around 4.4, in close agreement with the theoretical value 4.67.

Figure 101 shows the time dependence of the vertical velocity component measured at the center of the cell for increasing values of the Rayleigh number starting at $R/R_c = 270$.[19] As R increases, we go from a periodic regime (A) to one where the velocity is strongly perturbed and then relaxes toward the normal oscillatory regime (B). At even larger Rayleigh numbers (C) the behavior becomes more chaotic, resembling bursts of turbulence. As discussed previously, such a transition to chaos is called

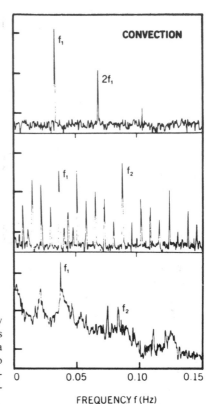

FIGURE 99.   As in Fig. 96, but for the velocity at the center of the Rayleigh–Bénard box. As the Rayleigh number (or heating) increases, a transition from periodic to quasi-periodic to chaotic evolution is again observed. (Reproduced by permission from Swinney and Gollub.[201])

intermittency. Other early studies with the Rayleigh–Bénard system were also important.[5,6,77,78,80,100,128,144,145]

In contrast to the nonlinear electrical systems and the Belousov–Zhabotinski reaction, where we presumably have a well-defined finite number of degrees of freedom, the Couette–Taylor and Rayleigh–Bénard systems are continuum hydrodynamic systems that, in principle, have an infinite number of degrees of freedom (in other words, each system is described by a system of *partial* differential equations whose phase space is infinite dimensional). When in the chaotic regime, however, only a few degrees of freedom are excited. Thus, even though irregular, the motion is described by some low-dimensional attractor. The dimension of this attractor has been estimated by Heslot *et al.*[103] to be around 2. In addition, Libchaber[125] presented results showing multifractal dimensions of attractors in the chaotic regime. Their studies, however, also showed that

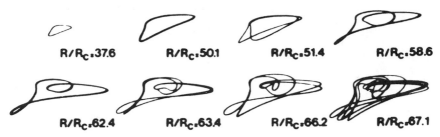

FIGURE 100.   Phase portraits (vertical temperature gradient versus horizontal temperature gradient) for increasing values of the Rayleigh number (R_c is a constant). This figure shows the period-doubling route to chaos. (Reproduced by permission of the American Physical Society from Giglio *et al.*[74])

the low-dimensional chaotic regime is quite narrow. As the Reynolds number increases, so does the dimension. For high Reynolds (or Rayleigh) number the systems pass from the chaotic state to a soft turbulence and then to a hard turbulence state where the dynamics are dictated by many degrees of freedom (high-dimensional attractors). Figure 102 shows ranges for the various regimes.

5. OTHER EXPERIMENTS

The experiments discussed are only several of many. Fingerprints of chaos have been found in systems such as a dripping faucet (Fig. 103) and mixing of fluids (Fig. 104), as well as in optical experiments,[9] biological experiments,[75,94] acoustics,[123] the Josephson system,[111] rigid-body systems and other engineering devices,[148] lazers,[40] and elsewhere.

In all, many of the phenomena predicted for dynamical systems have now been seen experimentally in a variety of systems. Recent success in estimating dimensions and Lyapunov exponents in the laboratory has closed the gap between the mathematical description of these systems and the reality of their observables.

Soon thereafter, scientists became curious. Is it possible that we could observe low-dimensional chaos in observables from "uncontrolled" experiments such as the weather, the stock market, or the brain? Were any secrets in Nature ready to unfold? Is the irregularity observed everywhere in Nature the result of low-dimensional strange attractors? If that were true, we could open new horizons in our understanding and describing of

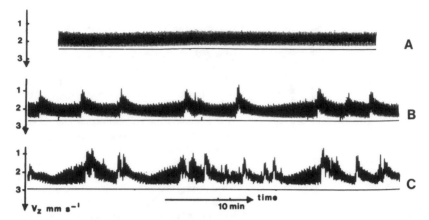

FIGURE 101.   Intermittency in the Rayleigh–Bénard system has also been observed. This figure shows the vertical velocity at the center of the box for increasing values of the Rayleigh number R. As R increases, we observed periodic behavior (A), intermittency (B), and finally chaos. (Figure courtesy of Dr. P. Bergé.)

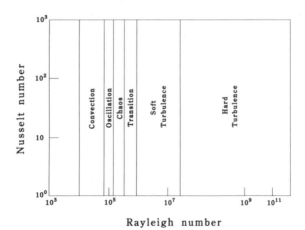

Rayleigh number

FIGURE 102.   Graph shows the Nusselt number (the ratio of the total heat transported across a Rayleigh–Bénard cell to the heat transported by conduction) as a function of the Rayleigh number. The various regions indicate the different scenarios (dynamics) taking place. As the Rayleigh number increases, we observe the onset of oscillatory instability followed by low-dimensional chaotic behavior and then by soft and hard turbulence where many degrees of freedom are excited.

DATA

MODEL

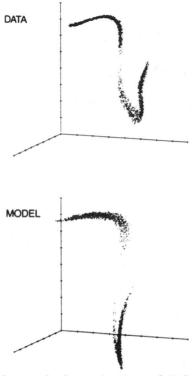

FIGURE 103. A simple but ingenious experiment was devised by Robert S. Shaw and his collaborators at the University of California at Santa Cruz. It consisted of allowing drops from a faucet to fall and measuring the time intervals between them. Thus, the observable is a variable that indicates the time interval t_1 between successive drops. (t_1 is the interval between the first and second drops, t_2 the time interval between the second and the third drops, etc.). If the drops fall regularly (at equal intervals), then in a t_n vs. t_{n+1} plot (or in a t_n vs. t_{n+1} vs. t_{n+2} plot) the result is simply a point indicating periodic behavior of period 1. If the drops fall at random, then such a diagram reveals a featureless structure. To their surprise for somewhat high flow rates, the reconstructed attractor in three dimensions looked like the top of this figure. The fact that a certain structure appears is an indication that the irregular falling of drops has a deterministic underpinning. The characteristic shape of the structure in the top can be thought of as a "snapshot" of some fold process. Such a structure compares favorably to a reconstructed attractor from an observable of the Hénon map shown in the bottom. Periodic regimes were observed from low flow rates. The authors (Crutchfield et al.[37]) comment that the changes between periodic and random-seeming patterns are reminiscent of the transition between laminar and turbulent fluid flow. (Reproduced by permission from Crutchfield et al.[37].)

many phenomena that seemed to be "too random" and thus usually treated in a statistical way.

6. DO LOW-DIMENSIONAL ATTRACTORS EXIST IN "UNCONTROLLED" PHYSICAL SYSTEMS?

In the last five to six years many studies have been published reporting evidence of low-dimensional attractors in weather, astronomy, biology, economics, epidemiology, and other fields. The evidence has included one or more of the following: finite correlation dimensions, positive Lyapunov exponents, and phase portraits. Table V presents selected studies together with their evidence.

FIGURE 104.   A beautiful example of chaotic and nonchaotic mixing in a Hamiltonian system has been presented by Ottino.[160] In this experiment a rectangular cavity is filled with glycerine, and two blobs of tracer that fluoresce respectively in green and in red are injected below the surface. Each side of the cavity can slide in a direction parallel to itself independently of the other sides. In the example in this figure the top side moves from left to right for a time interval and then stops. At this point the bottom side moves at the same speed and for the same time interval but from right to left. A complete movement of both sides constitutes a period. After five such periods the red blob has been stretched and folded several times while the stretching of the green blob has been insignificant. The red blob has been reduced to a pattern of folds within folds: it was placed in a region of chaotic mixing. (Reproduced by permission from Ottino.[160]) Other notable experimental studies are those by Behringer *et al.*[14] and del-Castillo-Negrete and Morrison,[41] which deal with Hamiltonian chaos and transport in quasi-geostrophic flows and Rössby waves. For a color reproduction of this figure see the color plates beginning facing page 148.

A complete study that follows the procedure outlined in Tables III and IV and is accompanied by phase portraits and physical interpretations has yet to be presented, even though some of them come quite close. Mainly, this is due to the fact that certain issues and developments (data requirements, choice of τ, etc.) arose or were refined after the appearance of many of these papers. In fact, the various algorithms used to estimate dimensions and Lyapunov exponents are still being refined. In addition, as discussed in Chapter 8, certain random processes and Hamiltonian systems (which do not possess attractors) were found to exhibit a finite correlation dimension. Arguably, these systems have no relevance to dissipative systems and thus to the time series sampled from physical systems. Nevertheless,

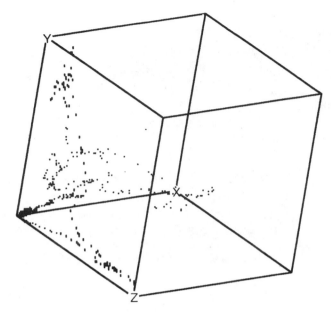

FIGURE 105.   Phase-space portrait of 5-sec rainfall data (Sharifi *et al.*[194] Figure courtesy of Dr. K. P. Georgakakos.)

as a result of these developments, the demands for convincing evidence are increasing accordingly, and many debates and/or controversies have arisen.

One of those controversies is the existence of low-dimensional attractors in weather and climate. The first study to report a low-dimensional attractor in climate was that of Nicolis and Nicolis.[154] This beautifully written paper, because of its clarity and style, sparked a great deal of enthusiasm among atmospheric scientists seeking a new theory. Other studies followed (see Table V). The work of Nicolis and Nicolis was criticized by Grassberger,[85] who argued that the number of points used was not large enough for a dimension of 3.1. Ruelle[185] also criticized Nicolis and Nicolis[154] and Tsonis and Elsner.[210,211] He argued, that according to his formula [Eq. (8.6)], both estimates were spurious. As demonstrated in Essex and Nerenberg,[51] however, Ruelle's comments (and his formula) are unfounded.

Tsonis *et al.*[215] offer an overview of all the latest developments concerning data requirements, proper choice of τ, and algorithm performance. In view of these developments they discuss the significance of the estimated

Heart Rate Dynamics

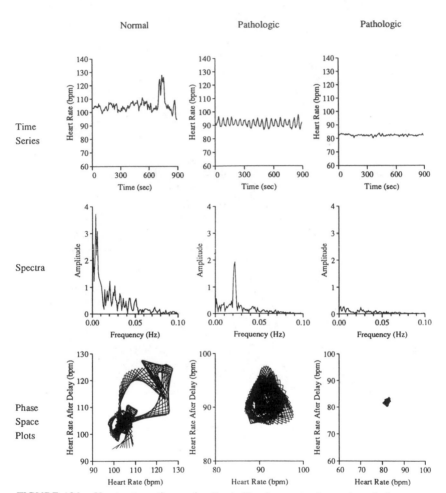

FIGURE 106.   Heart rate as time series (top), Fourier spectra (center), and phase-space plots (bottom). From left to right the plots correspond to healthy heart (chaotic dynamics), to a heart eight days before cardiac death (periodic), and to a heart 13 hr before cardiac arrest (nearly constant). (Figure courtesy of Dr. A. L. Coldberger.)

dimensions of weather and climate attractors. Some of their results are summarized in Fig. 111. The solid line is the function expressed by Eq. (8.4) or Eq. (8.5). The other symbols depict number of points used versus

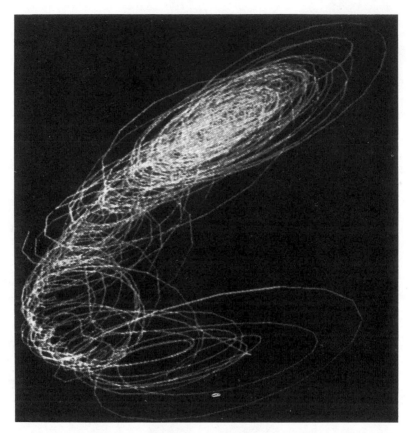

FIGURE 107.   Phase portrait made from electroencephalograms (EEGs). (Reproduced by permission from Dr. W. J. Freeman, 1991.) For a color reproduction of this figure see the color plates beginning facing page 148.

estimated correlation dimensions, as reported in Table V. Everything above or close to the solid line is considered as satisfying the data requirements. As we can see, although some claims cannot be substantiated, several studies might have used enough points.

Recently, while the arguments on the what is the right number of points are raging, Lorenz[138] considered a three-variable chaotic dynamical system and produced a model by taking seven linearly coupled copies of that system. The new system is described by 21 equations, and for a choice of the coupling coefficient its dimension is equal to 17.0. Lorenz applied

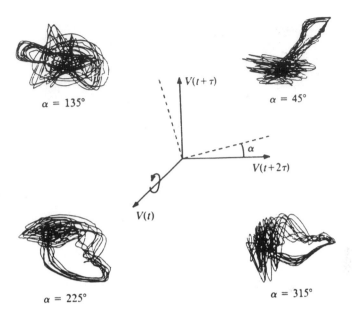

$$\alpha = 135°$$

$$\alpha = 45°$$

$$\alpha = 225°$$

$$\alpha = 315°$$

FIGURE 108.   Phase portraits of human epileptic seizure. First, the attractor is represented in a three-dimensional phase space. The figure shows two-dimensional projections after a rotation of an angle α around the $V(t)$ axis. The time series is constructed from the first (N = 5000 equidistant points and $\tau = 19\ \Delta t$). Nearly identical phase portraits are found for all τ in the range from 17 Δt to 25 Δt and also in other instances of seizure. (Reproduced by permission from Babloyantz and Destexhe.[11])

FIGURE 109.   Phase portrait generated from a time series representing the number of measles cases in New York City during 1928 to 1963. This is a 3D reconstruction using the method of delays and $\tau = 3$ months. Note the resemblance of the portrait to that of the Rössler system. (Reproduced by permission from Schaffer and Kot.[190])

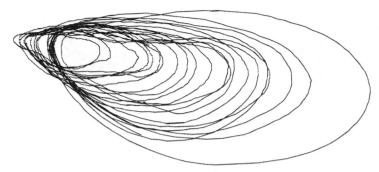

FIGURE 110.   Three-dimensional reconstruction of the attractor from the filtered sunspot data using the method of delays with $\tau = 10$ months. (Figure courtesy of Dr. M. D. Mundt.)

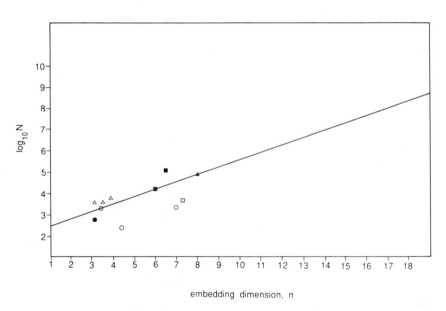

embedding dimension, n

FIGURE 111.   The solid curve is the function expressed by Eq. (8.4). The other symbols depict number of points used and attractor dimensions reported by various investigators. Everything closed or above the solid line can be considered as convincing evidence regarding the existence of low-dimensional attractors in weather and climate: ● Nicolis and Nicolis, [154] ○ Fraedrich, [63,64] ■ Essex et al., [49] □ Tsonis and Elsner, [210] ▲ Keppenne and Nicolis, [117] △ Sharifi et al. [194] (Tsonis et al. [215]).

TABLE V

Selected Studies That Have Reported Evidence for Low-Dimensional Attractors

	Type of data	Sample size N	Correlation dimension D_2	Lyapunov exponents	Phase portrait
Weather and climate					
Nicolis and Nicolis[154]	Oxygen isotope concentration	500	3.1	—	Yes
Fraedrich[63]	Pressure and sunshine	1,800	3.5–7.0	—	No
Fraedrich[64]	Oxygen isotope concentration	200	4.4	—	No
Essex et al.[49]	Geopotential	110,000	6.0	—	No
Tsonis and Elsner[210]	Vertical wind speed	4,000	7.3	—	No
Keppenne and Nicolis[117]	Geopotential	80,000	8.0	$\lambda_1 = 0.023$ $\lambda_2 = 0.014$	No
Sharifi et al.[194]	Rainfall	3,000	3.0–4.0	—	Yes (see Fig. 105)
Physiology					
Coldberger et al.[36a]	Heartbeat		—	—	Yes (see Fig. 106)
Rapp et al.[172]	Electroencephalograms (EEGs)	2,000	2.2	—	No

Reference	System	N	Dimension	λ_1	Chaos?
Freeman[70]	EEG (resting)	20,000	—	—	Yes (see Fig. 107)
Frank et al.[66]	EEG (epileptic seizure)	5,000	5.6	$\lambda_1 = 1.0 + 0.2$	No
Babloyantz and Destexhe[11]	EEG (deep sleep)	5,000	4.05	—	Yes
Babloyantz and Destexhe[11]	EEG (epileptic seizure)	5,000	2.05	$\lambda_1 = 2.9 \pm 0.06$	Yes (see Fig. 108)
Economics					
Sheinkman and LeBaron[196] (see also Brock and Seyers[25])		1,000	6.0		No
Epidemiology					
Schaffer and Kot[190]	Measles epidemics	420	2.55	—	Yes (see Fig. 109)
Astronomy					
Vassiliadis et al.[219]	Magnetosphere	5,000	3.6		No
Roberts et al.[173]	Magnetosphere	40,000	4.0		No
Shan et al.[193]	Magnetosphere	7,200	2.4		No
Mundt et al.[150]	Sunspot cycle	2,900	2.3	$\lambda_1 = 0.02$	Yes (see Fig. 110)
Pavlos et al.[162a]	Space plasmas	5,000–30,000	3.5–4.5	$\lambda_1 = 0.23$	Yes

the Grassberger-Procaccia algorithm, using 4000 values of a selected variable from the new system. He showed that, when N is not too large, (1) different selected variables can yield different dimension estimates, and (2) a suitably selected variable can sometimes yield a fairly good estimate. This suitable variable is strongly coupled to the rest of the variables of the system. This result is extremely interesting. If it holds for all chaotic systems, then the critical issue would not be the sample size of the observable but the observable itself!

Lorenz also demonstrated that if N is small the estimate of the dimension using a weakly coupled variable from the 21-equation system ought to resemble the estimate using a variable of the original three-variable system. Lorenz concludes that studies reporting on low-dimensional attractors in weather and climate are not altogether meaningless; they merely need to be reinterpreted. He states that, as suggested in Tsonis and Elsner,[211] the atmosphere might be viewed as a set of subsystems that are loosely coupled with each other. In this case what we measure is the dimension of a subsystem.

Another controversy is the existence of chaos in measles epidemics. The work of Schaffer and Kot[190] has been criticized because they used only 420 points. The estimated dimension of 2.55 is consistent with recent data requirements, and the phase space they produced is very interesting. However, a linear analysis by Schwartz[192] seemed to indicate that the data were merely a limit cycle with additive white noise. In Sugihara and May,[200] however, a prediction approach strongly supported the existence of low-dimensional chaos in the measles data (more on prediction in Chapter 10). Prediction also supports low-dimensional chaos in climate data similar to those used in Nicolis and Nicolis[154] (Elsner and Tsonis[47] see also Chapter 10).

It is true that some studies may now be questionable. On the other hand, some of the evidence is quite strong. Many studies have used enough points [according to Eq. (8.5)], and some of them have produced some stunning phase portraits of similar quality as those obtained from deterministic dynamical systems or "controlled" physical systems. It may be true, as many will argue, that it is impossible for the atmosphere (a system with infinite degrees of freedom) to exhibit a low-dimensional attractor. The possibility, however, that low-dimensional attractors exist in natural systems should not be dismissed simply because it does not "seem likely." On the other hand, it is the responsibility of the advocates of low-dimensional attractors to completely convince the disbelievers with solid analyses and enough physical interpretation of the results. The latter is particularly important. Presenting a value for the dimension of the attractor may not mean much anymore. More evidence is necessary. Next we discuss how nonlinear prediction might greatly fortify claims of low-dimensional chaotic dynamics.

CHAPTER 10

NONLINEAR TIME SERIES FORECASTING

Having gone through Chapters 8 and 9, you may justifiably get the feeling that subjective judgment may be required in order to determine the existence of an underlying chaotic attractor dictating the dynamics of an observable. Limitations of the algorithms, data requirements, and other problems discussed in these chapters often make the interpretation of the results a rather complicated issue. What can we possibly do in order to assist the search for low-dimensional attractors?

While we look for other approaches, we should not forget that chaos is deterministic. Chaotic systems obey certain rules. They have limited predictive power because of their sensitivity to initial conditions and because we cannot make perfect measurements (which require an infinite amount of information). However, before their predictive power is lost (i.e., for short time scales) their predictability may be quite adequate and possibly better than the predictive power of statistical forecasting. Figure 112 demonstrates the validity of the preceding statement. Consider the logistic equation $x_{n+1} = 4x_n(1 - x_n)$. We know by now that this equation generates chaotic sequences that are statistically indistinguishable from white noise. Since every value is uncorrelated to the previous or next value, the best forecast one can make is persistence (i.e., all future values will be equal to the present value).

We can graphically present the performance of persistence as follows. Consider an initial condition x_0 that produces a sequence $x_1, x_2, \ldots, x_n$ using the logistic equation. Find the square error $\epsilon_i = (x_0 - x_i)^2$, $i = 1, \ldots, n$. This gives the forecast error when we employ persistence. Repeat the procedure for many initial conditions, and calculate the probability that after one iteration ($i = 1$), ϵ_1 would be less than a value E [i.e., $P(\epsilon_1 < E)$]. Plot this probability as a function of E for increasing i's. The results

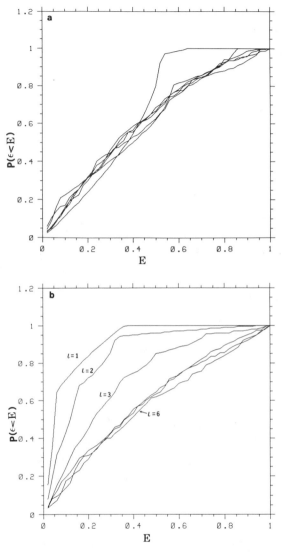

FIGURE 112. (a) Probability that the prediction error $\epsilon = (x_0 - x_i)^2$ is less than a value E as a function of E. Curves for the prediction error in i = 1, 2, 3, 4, 5, and 6 time steps into the future are shown. The prediction method is persistence. Note that all the curves are very similar, indicating that the predictive power of persistence is the same for any time step. (b) Same as in (a), but now we know the underlying mapping that takes present states into the future. Due to uncertainty in the exactness of the initial condition, the predictive power is eventually lost, but for very short times prediction is better than that in (a).

are shown in Fig. 112a. We see that see all these functions are quite similar, indicating that the predictive power of persistence in this case is the same for all time steps.

Let us again consider the initial condition x_0 and the generated sequence $x_1, x_2, \ldots, x_n$. We will assume that this is our "control" sequence. Then we perturb the initial condition x_0 by 10% and we get x_0'. We may think of x_0 as the exact initial condition and x_0' as the measured one. Starting from this initial condition and using the logistic equation $x_{n+1} = 4x_n(1 - x_n)$, we obtain the following sequence: $x_1', x_2', \ldots, x_n'$. Let us define the square error as $\epsilon_i = (x_i - x_i')^2$, $i = 1, \ldots, n$. By repeating the above procedure many times, we can again obtain the probability that ϵ_i would be less than a value E for increasing i's (i.e., time steps). The results for $i = 1, 2, 3, 4, 5$, and 6 are shown in Fig. 112b. According to this figure, after one iteration ($i = 1$) the probability that ϵ_1 is less than 0.4 is almost 1 (i.e., the error is not greater than 0.4). After two iterations ($i = 2$) the probability that ϵ_2 is less than 0.4 drop to 0.95. This drop in the probability is equivalent to the loss of predictability of our chaotic system after an iteration. However, it takes four iterations before the predictive power becomes equal to the predictive power of persistence. Therefore, the fact that we know the underlying determinism, even if we do not know the initial condition exactly, should help us improve short-term predictions.

Consequently, the basic philosophy behind nonlinear forecasting is the same as that behind estimating Lyapunov exponents: obtain from an observable the mapping that dictates where, in an n-dimensional phase space, the next point will be located.

1. GLOBAL AND LOCAL APPROXIMATIONS

Since we usually deal with dynamics in discrete time, we can assume that the underlying dynamics can be written as a map of the form

$$x(t + T) = f_T(x(t)) \tag{10.1}$$

where in phase space $x(t)$ is the current state and $x(t + T)$ is the state after a time interval T. Both f and x are n-dimensional vectors. The problem is then to find a good expression for f.

We attack this problem with a step-by-step example. Assume we are given the following sequence of numbers, $s(t)$, $t = 1, 18$: 0.41, 0.9676, $0.1254 \cdots$, $0.4387 \cdots$, $0.9849 \cdots$, $0.05921 \cdots$, $0.2228 \cdots$,

$0.69271 \cdots , \quad 0.85143 \cdots , \quad 0.5059 \cdots , \quad 0.9998 \cdots , \quad 0.0005682 \cdots ,$
$0.002271 \cdots , 0.009066 \cdots , 0.03593 \cdots , 0.1385 \cdots , 0.4775 \cdots ,$ and
$0.9979 \cdots .$ How can we obtain a prediction for $s(19)$?

Using the method of delays and $\tau = 1$, we can construct a 2D phase space with coordinates $x_1(t) = s(t)$ and $x_2(t) = s(t + \tau)$. When the dimension D of the attractor is known, the proper embedding dimension d should be consistent with Takens's theorem $(d \sim 2D + 1)$. Such a reconstruction will result in a sequence of points in phase space, $x(t)$. Figure 113 shows the resulting points. Point 1 $[x(1)]$ is the point with coordinates $(0.41, 0.9676)$, point 2 $[x(2)]$ is the point with coordinates $(0.9676, 0.1254 \cdots)$, etc. In this example it is easy to see that the function f_1 in $x(t + 1) = f_1(x(t))$ is quadratic. Thus, we expect that $x(t + 1) = a + bx(t) + cx^2(t)$. The task now is to determine the constants a, b, and c. Usually, since we have many more points than constants, we are given m equations in k unknowns with $m > k$. In our case $m = 17$, $k = 3$. The problem is overdetermined, and a solution does not usually exist. In fact, we want the problem to be overdetermined. If it is not, noise might prevent us from capturing enough dynamics.

For m points,

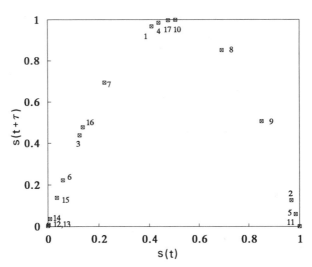

FIGURE 113. Two-dimensional phase-space reconstruction using the method of delays with $\tau = 1$. Coordinates $x_1(t)$ and $x_2(t)$ are defined as $x_1(t) = s(t)$ and $x_2(t) = s(t + \tau)$, where $s(t)$ is a the time series: $0.41, 0.9676, 0.1254 \cdots , 0.4387 \cdots , 0.9849 \cdots , 0.05921 \cdots ,$ $0.2228 \cdots , 0.69271 \cdots , 0.85143 \cdots , 0.5059 \cdots , 0.9998 \cdots , 0.0005682 \cdots ,$ $0.002271 \cdots , 0.009066 \cdots , 0.03593 \cdots , 0.1385 \cdots , 0.4775 \cdots ,$ and $0.9979 \cdots .$

$$x(2) = a + bx(1) + cx^2(1)$$
$$x(3) = a + bx(2) + cx^2(2)$$
$$\vdots$$
$$x(m) = a + bx(m-1) + cx^2(m-1)$$

We can express the system as

$$A = BC \qquad (10.2)$$

where

$$A = \begin{bmatrix} x(2) \\ x(3) \\ \vdots \\ x(m) \end{bmatrix} \quad B = \begin{bmatrix} 1 & x(1) & x^2(1) \\ 1 & x(2) & x^2(2) \\ \vdots & \vdots & \vdots \\ 1 & x(m-1) & x^2(m-1) \end{bmatrix} \quad C = \begin{bmatrix} a \\ b \\ c \end{bmatrix}$$

What we are looking for is C, but due to overestimation we cannot simply solve for $C = AB^{-1}$ (B is not invertible). In this case we seek to find a vector C that minimizes the error $e = \|BC - A\|^2$. We may think of it as a least-square linear regression, $y = a + bx$, where we are given many points (x_i, y_i) that do not fall on a straight line. In this case we seek the fit that will minimize the total error $\Sigma (y_i - (a + bx_i))^2$. As shown by Strang,[199] the least-square solution to $A = BC$ is the vector $C = (B^T B)^{-1} B^T A$. The least-square solution leads to the constants $a \sim 0$, $b \sim 4$, $c \sim -4$. Thus, our sequence is described by

$$x(t + 1) = 4x(t) - 4x^2(t)$$

or

$$x(t + 1) = 4x(t)(1 - x(t))$$

This is the logistic map with $\mu = 4$. Other ways to solve for C are based on factorization of the matrix B. The most popular method seems to be singular value decomposition (SVD), by which B is "decomposed" into Q_1, Σ, and Q_2 via the relation $B = Q_1 \Sigma Q_2^T$. If B is an $m \times n$ matrix of rank r, then Q_1 and Q_2 are orthogonal matrices of orders m and n, respectively, and their columns contain the eigenvectors of BB^T and $B^T B$.

The matrix Σ is an $m \times n$ diagonal matrix whose only nonzero entries are the r positive eigenvalues of $B^T B$ (which are also the r positive eigenvalues of BB^T).

The above fit is a *global* fit, and in a sense it takes care of our problem completely. Now we can use the mapping to predict, in principle, many time steps into the future by composing f with itself. However, we should expect that errors in approximation will grow exponentially with each composition. Thus, in practice we may have to fit a new function f for each time T. Moreover, global fits may not be as feasible as was the case with our example if f is complicated. Since chaotic dynamics do not occur unless f is nonlinear, for global fits we must be looking for nonlinear expressions for f. Polynomials are good representations, but, as Table VI explains, the number of parameters to be fit is equal to d^m, where d is the dimension of the phase space and m is the degree of the polynomial. Thus, fitting polynomials to obtain a global fit may become very impractical. In addition, since $\|y\| \rightarrow \infty$ (where y is a d-dimensional polynomial), as $\|x\| \rightarrow \infty$ polynomials have the disadvantage that they may not extrapolate well beyond their domain of validity.

A way to get around this problem is by representing f by the ratio of two polynomials.[171] Such an approximation is called *rational*. In general, it extrapolates better than polynomials, especially when the numerator and denominator are of the same degree (in this case the ratio remains bounded as $\|x\| \rightarrow \infty$). Other alternatives to polynomials are the *radial basis func-*

TABLE VI

When Fitting Polynomials, the Number of Parameters To Be Fit Is Equal to d^m, Where d Is the Dimension of the Phase Space and m Is the Degree of the Polynomial

Degree of polynomial	Embedding dimension d	Expression	Number of parameter, p	
	2	$y = a_1 x + a_2$	2	
$m = 1$	3	$y = a_1 x + a_2 z + a_3$	3	$p = d$
	4	$y = a_1 x + a_2 z + a_3 w + a_4$	4	
$m = 2$	2	$y = a_1 x + a_2 x^2 + a_3$	3	
	3	$y = a_1 x + a_2 x^2 + a_3 z$ $+ a_4 z^2 + a_5 y^2 + a_6 xz$ $+ a_7 xy + a_8 yz + a_9$	9	$p = d^2$
			Generalizing	$p \approx d^m$

tions and *neural networks*. Radial basis functions[170] are global interpolation schemes with good localization properties. As initially suggested by Casdagli,[28] they make good candidates in nonlinear modeling and forecasting. A simple expression for these functions is

$$R(x(t)) = \sum_{t'=1}^{N} \lambda_i \phi(\|x(t) - x(t')\|)$$

where t' ($t' < t$) is a label representing a point in a sequence length N and $\| \ \|$ is the Euclidean norm (the distance between two points).

The parameters λ_i are chosen to satisfy the condition $x(t' + T) = R(x(t'))$. In this form the radial basis functions depend on the distance between points only.

Neural networks provide another alternative for global fits. They are discussed in greater detail in Section 4.

In general, global fits provide good approximations if f is well behaved and not very complicated. A better approach is the *local* approximation introduced by Farmer and Sidorowich.[53-55] According to the local approximation, only the states near the present state are used to make predictions. Let us clarify this point by using Fig. 113. We take the last point, which represents $s(17)$ and $s(18)$. Point 18 [$x(18)$] represents $s(18)$ and $s(19)$. If we know where point 17 should go in the next step, we can obtain a prediction for $s(19)$. Very close to point 17 we find points 1, 4, and 10, with point 10 being the closest neighbor. In the past those points evolved to points 2, 5, and 11. The idea behind local approximation is that, instead of finding f by using all the data, we find f by using only the points in the neighborhood of point 17. In fact, since we only do this locally, we can be a little more flexible with our approximation.

We can start with the simplest example of local approximation, the first-order or nearest-neighbor approximation. According to this approximation, $s(19)$ is approximated to the point to which the nearest neighbor of $x(17)$ evolved. In our example, the nearest neighbor to point 17 is $x(10)$, which evolved to $x(11)$, whose coordinates are (0.9998, 0.0005682$\cdots$). Thus, a prediction for $s(19)$ is simply $s(19) \sim 0.0005682$. The true value for $s(19)$ is ~ 0.008080. The forecast error is obviously very small. Having a value for $s(19)$, we can now consider its nearest neighbor and obtain $s(20)$, and so on.

Such an approach was initially employed in practice by Lorenz,[136] who tried to predict the weather via analogs. According to the analog approach, to predict the weather for tomorrow we (1) look in the past and try to identify the closest weather pattern to that of today, and (2) assume that the prediction for tomorrow's weather will be same as the weather one day later to the closest weather pattern. Lorenz's results were not very encouraging because of the limited data set (he only used 4000 weather maps).

An improvement to the first-order approximation is the second-order or local linear approximation. Now we consider all points in the neighborhood of point 17 (i.e., points 1, 4, and 10), and we fit a linear polynomial between $x(1)$ and $x(2)$ (the point to which point 1 evolved), between $x(4)$ and $x(5)$, and between $x(10)$ and $x(11)$. For our example we thus have

$$x(2) = a + bx(1)$$

$$x(5) = a + bx(4)$$

$$x(11) = a + bx(10)$$

where

$$x(2) = \begin{bmatrix} s(2) \\ s(3) \end{bmatrix} \qquad x(1) = \begin{bmatrix} s(1) \\ s(2) \end{bmatrix} \qquad \text{etc.}$$

This can be written in matrix form as

$$A = BC$$

where

$$A = \begin{bmatrix} x(2) \\ x(5) \\ x(11) \end{bmatrix} \qquad B = \begin{bmatrix} 1 & x(1) \\ 1 & x(4) \\ 1 & x(10) \end{bmatrix} \qquad C = \begin{bmatrix} a \\ b \end{bmatrix}$$

The task now is to solve for the matrix C that provides the estimates for the parameters a and b of the linear polynomial. Then a prediction for $x(18)$ follows from the equation $x(18) = a + bx(17)$. Once we have a prediction for $x(t + 1)$, we may determine the neighborhood of $x(t + 1)$ and obtain a prediction for $x(t + 2)$, and so on. Such an approach is referred to as direct forecasting, and it requires constructing f at each time

T. An alternative is to obtain f for $T = 1$ and then iterate f to predict for $T = 2, 3, \ldots$. Such iterative forecasting is simpler, although it is not as accurate as direct forecasting unless f is known very well.[54,55]

Such a procedure raises the following question: What is the optimal number of neighbors? According to Farmer and Sidorowich,[53] for a point $x(t)$ in an n-dimensional phase space a simple criterion might be obtained by minimizing $\sum \| x(t) - x(t_i) \|$, where $x(t_i)$, $i = 1, \ldots, n$, are neighboring points and $t > t_1, > t_2 \cdots > t_n$. The second-order approximation has the disadvantage that it is linear, but it has the advantage that the number of free parameters to be fit, and thus the size of the neighborhood, grow slowly with the embedding dimension (see Table VI). Obviously, if we increase the order of the local approximation, we increase the accuracy. The problem is that there are so many nonlinear representations. Thus, finding the right representation is a difficult problem.

Figure 114 shows the prediction error E as a function of the prediction time T for the logistic map with $\mu = 4$ for the first-order, second-order, and third-order $[x(t + T) = a + bx(t) + cx^2(t)]$ approximation, respectively. The error E is the normalized root-mean-square error defined as $E = [\langle (x_p(t, T) - x_a(t + T))^2 \rangle]^{1/2}/[\langle (x - \langle x \rangle)^2 \rangle]^{1/2}$, where the angle brackets denote the average over all values and p and a stand for predicted and actual, respectively. Results from both direct and iterative forecasting are shown. As expected, the accuracy increases with the order of the local approximation.

If the representation f is known very well the error associated with direct forecasting is[54]

$$E \sim N^{-(m+1)/D} e^{(m+1)\lambda T} \qquad (10.3)$$

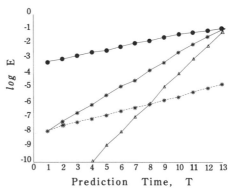

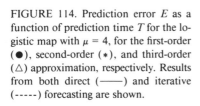

FIGURE 114. Prediction error E as a function of prediction time T for the logistic map with $\mu = 4$, for the first-order ($\bullet$), second-order ($*$), and third-order ($\triangle$) approximation, respectively. Results from both direct (——) and iterative (-----) forecasting are shown.

The error associated with iterative forecasting is

$$E \sim N^{-(m+1)/D} e^{\lambda T} \tag{10.4}$$

where D is the dimension of the attractor, N is the number of points in the time series, m is the order of approximation, λ is the largest Lyapunov exponent, and T is the prediction time. Thus, for $m = 0$, iterative forecasting is as accurate as direct forecasting, whereas for $m > 0$ iterative forecasting is better than direct forecasting. This might seem odd, since direct forecasting is "adjusted" every time and since there is no error accumulation due to iteration. That iterative predictions are superior may be explained from the fact that they make use of the regular structure of the higher iterates. It seems more natural to iterate, since the time series has been obtained by successive iterations. Note, however, that iterative forecasting will *not* always result in better predictions than direct forecasting. Employing neighborhoods is like looking at the structure of the attractor in many places. Such a procedure is connected to the underlying geometry of the attractor. That is why neighbors and neighborhoods are utilized in prediction and in approaches to compute dimensions[13,73] and Lyapunov exponents.[45,46,189]

2. EXAMPLES

The local approximation has been successfully used to improve predictions for laboratory data and for observables from physical systems. Figure 115 shows an experimental time series (top) obtained from Rayleigh-Benard convection in an He^3–He^4 mixture with Rayleigh number R/R_c = 12.24 and $D = 3.1$[53] On the bottom the normalized prediction error is shown as a function of the prediction time: LL stands for local linear, and GL stands for global linear. The global linear model is a linear autoregressive model we usually employ when we treat the data statistically. The number following LL or GL indicates the embedding dimension. The dashed straight lines represent the scaling dictated by Eq. (10.3). Obviously, for the correct embedding dimension there is no comparison between the local approximation and the autoregressive model.

Figure 116a shows a time series generated from the monthly number of cases of measles, x, in New York between 1928 and 1972. The time series indicates $x_{t+1} - x_t$ as a function of time. For this data, an attractor of a dimension between 2 and 3 has been suggested.[190] The time series from 1928 to 1946 was the training part used to make predictions from 1946 to

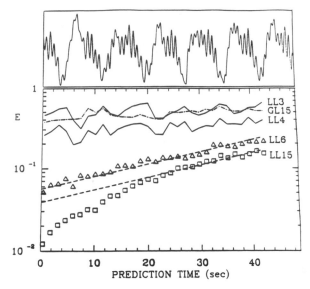

FIGURE 115. Result from applying the local approximation to laboratory data. The data (top) represent a time series obtained from Rayleigh-Benard convection. The dimension of these data is ≈3.1. In the bottom the normalized prediction error E for local linear (LL) and global linear (linear autoregressive, GL) is shown. The number following the initials indicates embedding dimension. The dashed lines are from Eq. (10.3). Obviously, nonlinear prediction is superior to linear prediction. (Reproduced by permission from Farmer and Sidorowich[53].)

1963 using a simple variant of the local approximation.[200] Figure 116b is a scatter diagram, with the coordinates being the observed and the predicted values for embedding dimension 6, time delay $\tau = 1$, and prediction time T_p = 1. With diagrams like this a perfect prediction will fall on the diagonal. Figure 116c shows the correlation coefficient of the data in 116b as a function of the embedding dimension. Note that for embedding dimension 5 or higher the correlation coefficient remains more or less constant. Thus, the optimal embedding is around 5 to 7, in agreement with the speculated dimension D = 2-3 and Takens's theorem. Figure 116d shows the correlation coefficient as a function of the prediction time T_p for embedding dimension 6 and τ = 1. Since the predictive power of a chaotic system falls with time, the fall of the correlation coefficient with T_p is consistent with the presence of chaotic dynamics in the measles record. If the record were uncorrelated noise, then the correlation coefficient would remain constant with T_p. For points connected by solid lines, the predictions are for the second half of the time series (based on a library of patterns compiled from the first half). For points con-

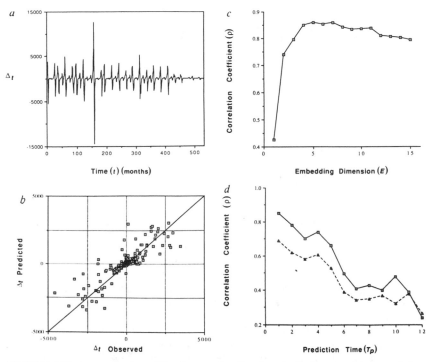

FIGURE 116. An application of local approximation (see text). Reprinted by permission from Dr. G. Sugihara and from *Nature* **344,** 734–741, copyright 1990, Macmillan Magazine, Ltd.)

nected by dashed lines, the forecasts and the library of patterns span the same time period (the first half of the data). The similarity between solid and dashed curves indicates that secular trends in underlying parameters do not introduce significant complications. The overall decline in prediction accuracy with increasing time into the future is a signature of chaotic dynamics as distinct from uncorrelated additive noise (see also Section 7). Other applications of local approximation include prediction of sunspots[30,150] and prediction of speech.[207]

3. INPUT–OUTPUT SYSTEMS

Consider an unknown system shown in Fig. 117. Up to this point we have considered methods that delineate the dynamics and make predictions for an observable (response) $x(t)$ of that system. Often we may know from theory that an input $u(t)$ drives the unknown system and thus influences $x(t)$. The philosophy behind the modeling of dynamical systems via the input–output system is to use both input and output in reconstructing the dynamics of the unknown system.[29,109]

In a similar way to that for an observable, we may attempt to model an input–output system based on a delay coordinate reconstruction via the general expression

$$x(t) = F(x(t - \tau), x(t - 2\tau), \ldots, x(t - k\tau), u(t),$$
$$u(t - \tau), \ldots, u(t - (l - 1)\tau))$$

Here k and l are the number of delays of the response and input, respectively, and F is a nonlinear function whose representation depends on the system. The same functional forms described previously are suggested here as well. From all these forms it is suggested to use a local linear model:[30]

$$x(t) = b_1 x(t - \tau) + b_2 x(t - 2\tau) + \cdots + b_k x(t - k\tau)$$
$$+ a_0 u(t) + a_1 u(t - \tau) + \cdots + a_{l-1} u(x - (l - 1)\tau)$$

From here on the method proceeds in a way identical to that discussed in the case of local linear approximation. We simply have to solve for the coefficient array C in the equation $A = BC$, where A is an array containing the response values $x(t)$, and B is an array representing sets of lagged input and output values from the training data set.

A nice example of an input–output system is climate. One output of such a system is the known ice volume (or oxygen isotope) records, which were the first records used in the search for low-dimensional attractors in climate.[154] The input of such a system could be the solar insolation, which can theoretically be obtained as a function of time from celestial mechanics.[20] Solar insolation relates to the amount of solar energy received on

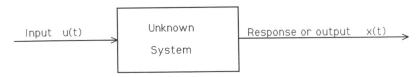

FIGURE 117. The response of an unknown system is often a result of some input that drives the system.

the top of the atmosphere and, therefore, regulates climate. The input–output time series are shown in Fig. 118.

Figure 119 shows prediction of the ice volume time series using the input–output approach. For the purpose of comparing predictions with available data, the predictions start 150 kyr before the present and continue for 100 kyr after the present. The solid line is the observed data, and the squares represent iterated predictions. Interestingly, according to the prediction (which works very well, as we can see from the comparison between observed and predicted values), the earth would cool for the next 40,000 years!

Other applications of the input–output approach include studies in mechanical vibrations, such as analog Duffing oscillator and a beam moving in a double potential well.[109]

4. NEURAL NETWORKS

We introduce the philosophy behind neural networks by presenting a highly simplified example, which is a modification of an example pre-

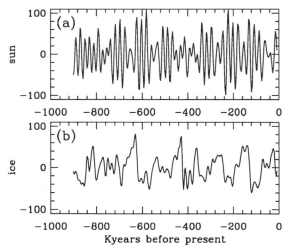

FIGURE 118. Input time series (relative solar insolation at 60° north latitude) and output (relative global ice volume based on oxygen isotope analysis of ocean core V 28-239). Solar insulation is related to the available solar energy, which drives the climate system whose output is the global temperature of the planet (or ice extent). (Figure courtesy of Dr. M. Casdagli.)

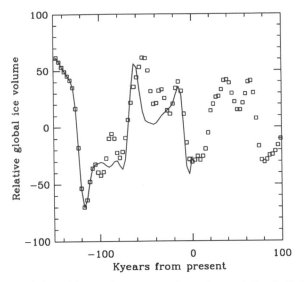

FIGURE 119. Prediction of the next ice age, assuming no human-induced effects. Shown are the ice age data up to the present (−), and the iterated predictions (□) starting from about 150,000 years ago and continuing into the future. (Figure courtesy of Dr. M. Casdagli.)

sented by Owens and Filkin.[161] Consider global precipitation over the past five years, with our interest focused on predicting the precipitation for 1990. Under such an arrangement we say that we have one *training pair* consisting of the five *inputs* $p[i]$ and a single *output node* Q. The relationship between the inputs and the output is shown in Fig. 120. Such a figure is often referred to as the *architecture* of the network. For this example it consists of two layers, an *input layer* and an *output layer*. The five inputs can be thought of as a five-component state vector, with the value of each component given as the amount of precipitation for the year. Usually the inputs are scaled to the range $0 \le p[i] \le 1$. The value P is constructed as the inner product given by the sum of the inputs multiplied by their corresponding connection weights $w[i]$:

$$P = \sum w[i]p[i]$$

The summation in the equation is over the five inputs. The output Q is obtained by passing the inner product P through a nonlinear function $f(x)$, sometimes called the squashing function. This function gives the

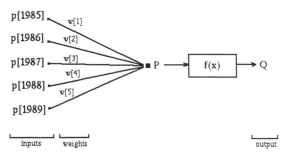

FIGURE 120. Schematic (architecture) of a two-layer neural network. Each of the five inputs has a value $p[\ ]$ corresponding to the amount of precipitation for that year. The output has a value corresponding to the amount of precipitation for 1990. The weights ($w[\ ]$) indicate the relative strength of connections between inputs and output. The input values are combined with the weights by an inner product to give a value P. Inputs are taken to outputs by using a nonlinear squashing function such as tanh. (Elsner and Tsonis.[47])

neural networks their nonlinear character. The squashing function has limits $0 \le f(x) \le 1$, which guarantees that the output Q is limited in range regardless of the value of P. For this example, we have a single input–output training pair denoted (P, Q). All the connection weights are then varied to minimize the squared error, calculated as the difference between the network's predicted output and the actual value.

In our simple one-pair, two-layer network in Fig. 120 the training pair associates one specified set of inputs, for example $p[1985]$, $p[1986]$, $p[1987]$, $p[1988]$, $p[1989]$ with a single output Q. The error to be minimized is the squared difference between the actual value $p[1990]$ and the network value Q:

$$E = (p[1990] - Q)^2$$

The weights are changed by first finding the gradient of E with respect to $w[i]$ and then adjusting $w[i]$ to force E toward smaller values. This is accomplished with the help of a forward Euler integration scheme:

$$w[i]^{n+1} = w[i]^n + \eta \, \Delta w[i]$$

where $\Delta w[i] = -\delta E / \delta w[i]$. The term $w[i]^n$ represents the weight at iteration number n, and η is the learning rate.

This is analogous to finding the root of a polynomial by using the generalized Newton method, where convergence to a root is achieved by successive evaluations of the function and its derivative. For more than one training pair, the equation is generalized by summing over all training pairs. Training occurs in discrete iterations, with each requiring one presentation of all training pairs to the network. The network "learns" by presenting the (P, Q) pair sequentially with a number of training pairs relating the values of the input to a corresponding value of the output. The learning rate η must be small to ensure convergence of the integration scheme. However, a stiff integration technique can greatly improve the learning rate.[161]

In general, neural network programs are built around the concept of adjustable weights that take inputs to outputs. Each weight carries information that indicates how strongly the input is connected to the output. We can now formally define neural networks. A simple neural network model can be written as

$$x[i] = x(t - i\tau') \qquad z = f(\sum w[i]x[i])$$

where $f(x)$ is a nonlinear sigmoidal function, such as the hyperbolic tangent, where the $x[i]$'s are the inputs, which in effect form the coordinates of a state-space. The parameter τ' is usually set to unity, but can assume other values as well. Note that the dimensionality of this space is equal to the number of inputs i. Thus, in the example the dimensionality is 5. If we had used only four years prior to 1990, then the dimensionality would have been 4. In a way the number of inputs defines an embedding dimension. This embedding dimension, however, may not be the same as the embedding dimension of the Takens theorem. While intuitively it seems that those two embedding dimensions should be related, up to this point this relation has not been established. In practice, one uses as many inputs as it takes in order to obtain the desired results. Therefore, even though the *a priori* knowledge of the dimension of the underlying attractor may suggest a first guess for the number of inputs, its exact value is not required.

Starting with arbitrary values for the weights ($w[i]$), we calculate an output (z) and compare it with the actual value $x(t_i + T)$. The squared error between the model output and the actual value

$$E = [x(t_i + T) - z]^2$$

is subsequently used to change the weights. This is done by first calculating the derivatives of the error with respect to all the weights ($\delta E/\delta w[i]$).

Then, if increasing a given weight leads to more error, the weight is adjusted downward. Otherwise, if increasing the weight leads to less error, the weight is adjusted upward. Since information about the error at the output layer is used to modify the weights at the input layer, the method is called back propagation.[187] The procedure is continued until all the weights and the error settle down to below some prescribed tolerance. Commonly the initial weights are chosen as uniformly distributed random numbers; however, if prior information exists, a better initial guess can be made.[222]

Often when the system of interest is sufficiently complex (involving many degrees of freedom) a second layer, called a *hidden layer,* is added to the network (see Fig. 121). The multilayer neural network can then be written as

$$x[i] = x(t - i\tau')$$
$$y[j] = f(\sum w[i,j]x[i])$$
$$z = f(\sum w[j]y[j])$$

Connection weights are specified between input and hidden values ($w[i, j]$) and between hidden values and the output ($w[j]$). The weight-modifying scheme is applied in the same manner to the multilayer network.

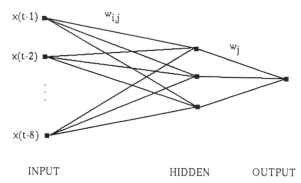

INPUT   HIDDEN   OUTPUT

FIGURE 121. Architecture of the neural network used in the examples. The single output corresponds to the fact that we are predicting one step into the future. The number of hidden values is set at 3. Results from numerous trial runs indicated that adding more hidden values did not significantly improve the networks prediction capabilities. The number of inputs varied for the different examples. Again, however, the model was not sensitive to small changes in the number of inputs used. The values of the inputs nodes are lagged values of the time series. (Elsner and Tsonis.[47])

We may summarize all this as follows. With the help of known outputs the network, initially set to a randomly chosen state, modifies its structure (changes the weights) in such a way as to improve its performance over the training set. If the network architecture is rich enough (i.e., sufficient number of both inputs and layers), this procedure eventually leads the network to a state in which inputs are correctly mapped to outputs for all chosen training pairs.[116]

Neural networks, along with the other dynamical state-space models, are phenomenological in that they assess the qualitative characteristics of the underlying system's dynamics. Short-term predictions are based on that knowledge, without providing a physical understanding of the mechanisms that might be operating within the system. However, successful predictions with such models can lead to useful hypotheses concerning, for example, why certain inputs are associated with stronger connection weights compared with others, which can readily be interpreted as a hypothesis concerning the physical nature of the system. Those that appreciate such views may also appreciate the views of Wiener.[223,224]

5. EXAMPLES

In this section we present three examples[47] showing the effectiveness of using a neural network for making predictions on time series data. Each example uses a different data set. For the first example we use data generated artificially; for the second example we use data generated from a controlled laboratory experiment; and for the third example we use data observed in nature. The neural network architecture we employ for each example consists of three layers: input, hidden, and output. Learning is achieved by using the method of back propagation as discussed in the previous section. Training is performed on the first part of the time series with subsequent predictions made on the remaining values. For each example, the number of outputs is set to 1, the number of hidden values is set to 3, while the number of inputs depends on the individual example. Numerous trial runs indicated that the accuracy of prediction was not sensitive to small changes in the number of inputs or hidden values. The single output represents some future value of the time series we wish to predict.

The neural network architecture is shown in Fig. 121. The inputs are the components of a reconstructed n-dimensional state-space consisting of successive time-delayed values of the series. The method is similar to the one used by Perrett and van Stekelenborg[165] to predict annual sunspot numbers. We represent the series as $x(t_i)$, where $i = 1, 2, \ldots, L$. With τ' = 1 and an eight-dimensional phase space (i.e., eight inputs) beginning with the first value of the time series, the first set of inputs is $[x(1), x(2), \ldots, x(8)]$ and our output is $x(9)$. Similarly, the second set of inputs is $[x(2), x(3), \ldots, x(9)]$, and the output we are trying to predict is $x(10)$. Training continues over all training pairs (set of inputs, output) for several thousand iterations.

For the first example, we generate a time series by numerically integrating the Lorenz system, using a fourth-order Runge-Kutta integration scheme and constants $a = 16.0$, $b = 120.1$, and $c = 4.0$. The time series of convective motion (x component of the system) after all transients (10^4 iterations) have diminished is shown in Fig. 122a. Positive values indicate upward motion in the fluid. We take 1000 values from the time series, train the network on the first 500 values, and make predictions on the last 500 values. Results are thus based on sample sizes of ~ 500. The number of inputs in the network was 8. Results of the neural network at predicting one step into the future (points) compared with the actual values (solid line) are given in Fig. 122b. The normalized root-mean-square error (RMSE) between the actual and predicted values is 0.072, where zero implies a perfect forecast. Clearly, the network is capable of capturing the underlying chaotic dynamics of the system (see Frison[72]).

To assess the predictive ability of the neural network against that of a standard statistical model, we fit the first half of the time series, using an optimal autoregressive (AR) model, and then compare predictions on the second half of the series from both models. For the AR model the time series is viewed as a single realization of a stochastic process that is taken to be stationary and having a Gaussian distribution. For model selection we employed the Bayesian information criteria as outlined in Katz[115] and determined that the optimal order of the AR model for the time series is 12.

Comparisons between the neural network and AR models are made by quantifying how the prediction accuracy (skill) decreases as predictions are made further into the future. To do this we predict one step into the future and use this predicted value as one of the lagged inputs for the next prediction two time steps into the future. Doing this successively allows us to compute the correlation coefficient between actual and predicted values as a function of prediction time, where prediction time is given as discrete time steps into the future. This procedure is followed for both the neural network model and the optimal AR model.

Results are shown in Fig. 123. For the first few steps into the future,

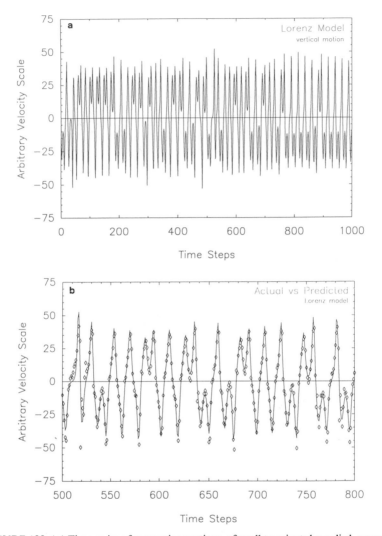

FIGURE 122. (a) Time series of convective motions, after all transients have died, generated by numerically integrating the Lorenz system using a fourth-order Runge-Kutta scheme. The time axis is in integration steps, and the magnitude of convection is on an arbitrary velocity scale. The series displays chaotic oscillations. (b) Comparison of the actual time series (solid line) with a neural network prediction (points). The number of inputs in the neural network is eight. The actual time series represents a part of a novel portion (second half) of the convective signal. Predicted values correspond quite well with actual values. (Elsner and Tsonis.[47])

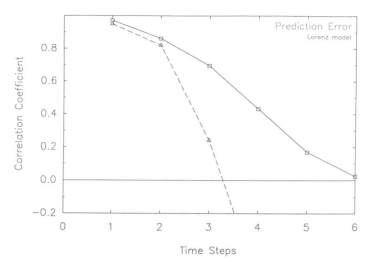

FIGURE 123. Correlation coefficient between actual values and predicted values and predicted values as a function of prediction time for the convective motions using a neural network model (solid line) and an optimal autoregressive model (dashed line). Prediction time is given as discrete time steps into the future. A correlation coefficient of 1 corresponds to perfect prediction. The neural network model clearly outperforms the autoregressive model. (Elsner and Tsonis.[47])

predictions from both models are good and the difference between the two models in terms of predictive skill is small. In contrast, the neural network (solid lines) makes significantly better forecasts than does the AR model (dashed line) as prediction time increases. Predictive skill on a nonuniform chaotic attractor will vary in time.[152] However, by using the same segment of the attractor to compare the models as was done here, we ensure a fair comparison. We note that the AR model is essentially a linear model and therefore incapable of capturing the inherent nonlinear nature of such a record. Since the signal is chaotic, we cannot hope to make accurate predictions too far into the future with any model. We see that the predictive skill of the neural network also drops to near zero after a relatively short time.

We next turn our attention to data generated from a controlled fluid dynamics experiment. The data were recorded from a rotating differentially heated annulus of fluid. The experiments were performed at the Geophysical Fluid Dynamics Institute (GFDI) to study the transition to turbulence in fluids. The experiment from which the data were taken is described in detail by Pfeffer et al.[166,167] For our purposes it is sufficient to say that the data recorded in time series represent temperatures in degrees Celsius (°C) at a single location near mid-depth in the fluid. The temperature contrast from the inner wall to outer wall of the annulus is held constant at 10°C. Sampling rate is once every two rotations, with each rotation analogous to one sidereal day.

As before, we take 1000 values from the time series, train the neural network on the first 500 values, and make predictions on the last 500 values. The complete record is shown in Fig. 124a, and a forecast one step into the future is shown in Fig. 124b. The normalized RMSE between actual and predicted values one step into the future is 0.065, indicating very good predictions. Shown in Fig. 125 is the correlation coefficient between actual and predicted values as a function of prediction time for both the neural network (solid line) and an optimal fourth-order AR model (dashed line). As seen previously, predictions from both models are good for the first few time steps into the future, and the differences between the two models in terms of predictive skill are small. After that, however, the neural network clearly outperforms the linear AR model. Here the number of inputs was 75.

For the third example we take a time-series record of sea surface temperatures in degrees Celsius constructed by proxy, using deep-sea ice core records of oxygen isotope concentrations. Data are available for approximately 1700–700 kyr before the present at a sampling rate of 2 kyr for a total of 498 values.[182] Similar records have been used in climate research. The complete time series is shown in Fig. 126. We train the neural network by using eight input values on the first 400 values and make predictions on the remaining 98 values. The RMSE between actual and predicted values one time step into the future is 0.170, indicating some skill. For comparisons we again employ an optimal fourth-order AR model and compare correlation coefficients between actual and predicted values as a function of prediction time for both models (Fig. 127). As with the previous two examples, the neural network forecast demonstrates considerably more skill than does the forecast using an AR model, especially after the first few time steps. This result supports earlier evidence of deterministic chaos in climate.[154]

6. USING CHAOS IN WEATHER PREDICTION

The examples discussed in the previous sections deal with time series forecasting. Ideas from chaos theory, however, can be extended and em-

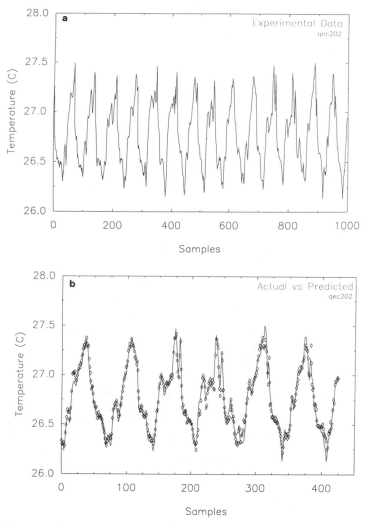

FIGURE 124. (a),(b). Same as Figure 122(a),(b), except the time series represents temperature (°C) taken from a rotating differentially heated fluid in an annulus. The record is taken at mid-depth in the fluid. The time axis is given in number of rotations times 2. The experiment was performed at the Geophysical Fluid Dynamics Institute. The number of inputs in the model is 75. The neural network makes excellent predictions one step into the future. (Elsner and Tsonis.[47])

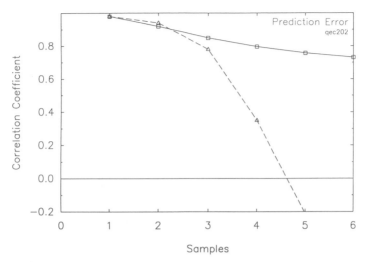

FIGURE 125. Correlation coefficient as a function of prediction time for the laboratory experiment data. The solid line represents predictions made with the neural network, and the dashed line represents predictions made with an optimal autoregressive model. Both models do well in the short term; however, the neural network clearly outperforms the linear autoregressive model as prediction time increases. (Elsner and Tsonis.[47])

ployed in various other situations. An excellent example is the possibility of using chaos to aid weather forecasting. The numerical models that are used in weather forecasting are the general circulation models (GCMs), which are based on the Navier-Stokes equations, the continuity equation, and other equations, taking into account thermodynamics, radiation, etc. Weather is presented in maps that show the pressure patterns over the globe at the surface as well as at selected levels in the vertical. Given an input field, the models can provide a 1-day, 2-day, . . . , n-day forecast.

Figure 128a shows a 5-forecast for April 30, 1990, made using the observed weather on April 25, 1990, as an input. Figure 128b shows a 4-forecast for April 30, 1990, made using the observed weather on April 26, 1990, as an input. The idea is that the April 26 input can be viewed as a small fluctuation of the input on April 25 (i.e., the large-scale weather does not change significantly over a period of 1 day). If the weather during this period is in a region in the weather attractor where the expansion of the trajectories is not very great, then the two forecasts should look very similar and, thus, could be trusted. If the two forecasts are very different, then the system is at a location where nearby states diverge very quickly. In this case the forecasts cannot be trusted. In our example both

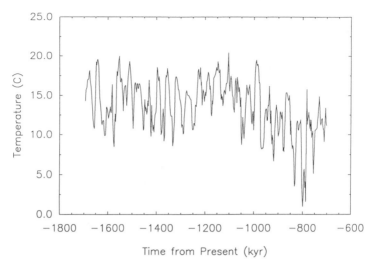

FIGURE 126. Time series of proxy sea surface temperatures in °C for the period 1700–700 kyr before the present at intervals of 2 kyr. The length of the record is 498 values. Similar records have been used in the study of climate dynamics. (Elsner and Tsonis.[47])

forecasts display substantial agreement in the large-scale features. Thus, the forecasts could be trusted. These ideas have been exploited by Dr. Kalnay and her colleagues at the National Meteorological Center (NMC) in Camp Springs, Maryland.[221]

Another area could be model comparison. For weather forecasting many models are used. These models differ in details in the physics, initialization, various parameterizations, etc. Because of the differences, their forecasts differ. Chaos may provide a way to explore ways to choose one of the forecasts for a particular area. For example, produce a global distribution of Lyapunov exponents (or some relevant measure of expansion of nearby states) for each model and simply follow the model that shows the smaller expansion rates over the area of interest.

7. CHAOS AND NOISE

The aforementioned results bring up an important point, which was indirectly made in Figs. 112a,b. Prediction can be used to confirm (or just identify) underlying chaotic dynamics. After all, determinism and pre-

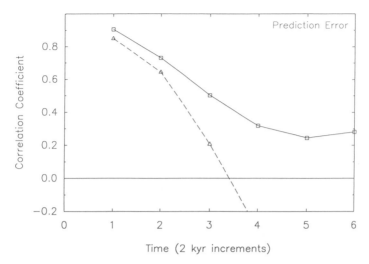

Time (2 kyr increments)

FIGURE 127. Correlation coefficients as a function of prediction time for the proxy sea-surface temperature data. The solid line represents predictions made with the neural network, and the dashed line represents predictions made with an optimal autoregressive model. The number of inputs in the network is 8. Both models do well in the short term; however, the neural network clearly outperforms the linear autoregressive model as prediction time increases. A rapid decrease of predictive skill with prediction time is characteristic of a chaotic signal. (Elsner and Tsonis.[47])

dictability are equivalent. Do not get confused here: unpredictability of chaotic systems is due to errors in the initial conditions, not to the absence of determinism. For short times, the underlying determinism would, if known, result in better forecasts than standard statistical techniques.

Since the predictive skill of uncorrelated noise is the same for any prediction time, nonlinear prediction could offer a way to distinguish between a chaotic signal and a noisy signal. Figure 129 demonstrates how.

After training the neural network on the first half of the signal composed of a sine wave plus noise, we make predictions on the second half, and, as with the Lorenz system, we compute the correlation coefficient between actual and predicted values as a function of prediction time. The dashed horizontal line in Fig. 129 is the result of this procedure. The independence of predictive skill with prediction length is in sharp contrast to the rapid decrease of predictive skill for the chaotic signal from the Lorenz system (solid line). The differences suggest that predicting time series using neural networks or any other nonlinear approach provides a method for differentiating additive noise from deterministic chaos.

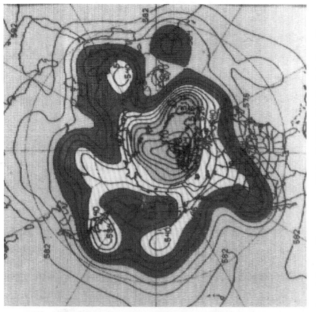

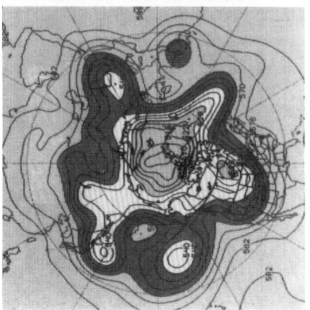

FIGURE 128. (a) Predicted weather map for 30 April 1990, using as an input the weather map on 25 April 1990. (b) Predicted weather map for 30 April 1990 using as input the weather map on 26 April 1990. The weather on April 26 might be considered as a small fluctuation of the weather on April 25. If these two states are in a region in the weather attractor where the divergence of nearby trajectories is small, then the two predictions for April 30 would be quite similar, and, therefore, they could be trusted. If the forecasts are very different, then the divergence of nearby states is great, and predictability would be lost fast. In this case the predictions for April 30 cannot be trusted. In our example the two predictions are very similar. Therefore, the prediction for April 30 should be trusted. Indeed, the observed weather on April 30, 1990 (map not shown) was quite similar to the predicted one. (Reproduced by permission from Dr. Kalnay of the National Meteorological Center and from *Weatherwise*, Heldref Publications.) For a color reproduction of this figure see the color plates beginning facing page 148.

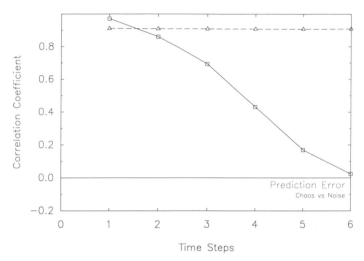

FIGURE 129. Correlation coefficient between actual and neural-network predicted values as a function of prediction time for the Lorenz system (solid line) and for a signal consisting of a sine wave plus noise (dashed line). The rapid drop in magnitude of the correlation coefficient with prediction time is characteristic of chaotic signals. In contrast, the independence of predictive skill with prediction time of the sine wave–plus–noise signal demonstrates that the neural network model is capable of distinguishing between additive noise and chaos. (Elsner and Tsonis.[47])

Predictions on time series with additive noise appear to have a fixed amount of error, regardless of how far into the future one tries to predict. On the other hand, prediction accuracy on chaotic time series degrade as one tries to predict too far into the future (see also Sugihara and May[200]). This suggests that it might be possible to quantitatively compare the rates of degradation in prediction skill to obtain an indication of just how chaotic a system is. In fact, Wales[220] presented a method according to which the largest Lyapunov exponent is estimated from the decay of the correlation coefficient with prediction time.

8. NONLINEAR PREDICTION AS A WAY OF DISTINGUISHING CHAOTIC SIGNALS FROM RANDOM FRACTAL SEQUENCES

We have seen how nonlinear prediction can be used to distinguish chaotic signals from periodic with additive noise signals even for limited data sizes. Nonlinear prediction can also be used to distinguish chaotic signals from purely periodic signals. For purely periodic signals the predictions will be perfect for any prediction time step. We have also demonstrated that in all cases linear statistical approaches, even though they might exhibit similar signature to the signature of nonlinear prediction, provide poorer predictions.

But how does nonlinear prediction perform when it comes to autocorrelated random sequences? Can nonlinear prediction distinguish, for example, chaos from random fractal sequences (fBm's)? The answer is yes, but it is not immediately obvious, because fractional Brownian motions exhibit a decrease of the correlation between predicted and actual values with prediction time. This decrease, however, obeys distinctly different rules.

Preliminary results of Tsonis and Elsner[214] indicate that nonlinear prediction is not as likely to be fooled by fBm's. If the predicted value at some t is assumed to be a random variable x_t and the actual value a random variable y_t, then Pearson's correlation coefficient between these two distributions, $r(t)$, ranges in magnitude between 0 and 1 for uncorrelated and identical distributions, respectively. The correlation coefficient is defined as

$$r(t) = \frac{\langle x_t y_t \rangle - \langle x_t \rangle \langle y_t \rangle}{\sigma(x_t)\sigma(y_t)} \qquad (10.5)$$

where the angle brackets denote the average over a series of predictions and σ denotes the standard deviation. For stationary chaotic signals the correlation coefficient can take the form[220]

$$r(t) = 1 - \frac{s^2(0)e^{2Kt}}{2\sigma^2(y_t)}$$

where $s(0)$, $\sigma(y_t)$, and K are certain positive constants. For stationary processes $\sigma(y_t)$ is considered independent of the prediction time t. The equation dictates that the correlation coefficient should exponentially decrease with prediction time. This agrees with the fact that in chaotic systems nearby trajectories diverge exponentially.

If x is an fBm, then we know (Chapter 4) that $x_t = N(0, \propto t^{2H})$ [i.e., $\langle x_t \rangle = 0$, $\sigma(x_t) \propto t^H$] with $0 < H < 1$, and that $cov(x_{t1}, x_{t2}) = \min(t_1, t_2)$. Therefore, it follows that $cov(x_t, y_t) = \langle x_t y_t \rangle$. In nonlinear prediction it is commonly assumed that $y_t = f(\sum_j w_j x_{t-j}) = f(z_t)$, where f is some

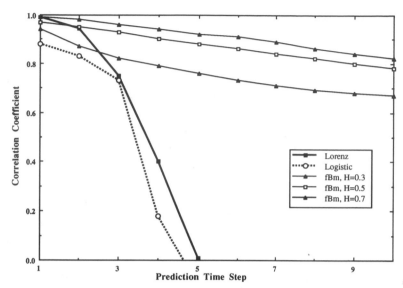

FIGURE 130. Nonlinear prediction can be used to distinguish chaotic signals from random fractal sequences. The figure shows the correlation coefficient between predicted and actual values as a function of the prediction time step for three fBm's with H equal to 0.7, 0.5, and 0.3, for the x coordinate of the Lorenz system, and for the logistic map $x_{n+1} = 4x_n(1 - x_n)$. The integration step for the Lorenz system was 0.03. The prediction method was neural networks. For each example we take 1,000 values from the corresponding time series, train the network on the first 500, and make predictions on the last 500 values. Thus, correlation coefficients are based on sample sizes of about 500. Note that for the Lorenz time series the prediction time step is the integration step. The training (learning) algorithm used was the back-propagation. A feedforward, three-layered network was used for all time series. For the fBm's we used a 3–8–1 network (i.e., 3 inputs, 8 hidden layers, 1 output). For the Lorenz system and the logistic map we employed an 8–3–1 network. While a large number of different network architectures can be considered, the networks employed here were the best of those tested, providing fast convergence and high correlations. The results show a decrease of the correlation coefficient with prediction time in all cases. Note, however, the differences in the scaling between chaotic signals and fBm's.

nonlinear function ranging between 0 and 1 and w_j's are certain coefficients. Expansion of f in a Taylor series gives

$$\langle x_t y_t \rangle = \mathrm{cov}\left(x_t, f(0) + z_t f'(0) + z_t^2 \frac{f''(0)}{2} + \cdots\right)$$

$$= \mathrm{cov}(x_t, f(0)) + f'(0)\mathrm{cov}(x_t, z_t) + \frac{f''(0)}{2}\mathrm{cov}(x_t, z_t^2) + \cdots$$

Considering the problem at hand, the equation yields an expression for the correlation coefficient of the general form $r(t) = F(t^{g(H)})$, which does not include exponential terms but only power-law terms. This is a direct consequence of the power laws involved in the characterization of fractional Brownian motions and it points out that significant differences in the scaling of the correlation coefficient with prediction time should be expected between chaotic signals and fBm's.

These scaling differences are highlighted by computed simulations. Figure 130 shows the correlation coefficient as a function of the prediction time step for the x coordinate of the Lorenz system, for the logistic map with $\mu = 4$, and for various fBm's. The curves that correspond to the chaotic signals show a significant fall-off whereas the curves that correspond to the fBm's exhibit a much slower rate of decrease. In fact, semi-log and log-log graphs of $1 - r(t)$ versus t indicate that for very short prediction time steps the best fit for the chaotic signals is an exponential fit whereas the best fit for the fBm's is a power-law fit. It is cautioned, however, that the correct type of scaling may not be delineated if too few points are used. In such cases different realizations of the same observable may not yield the same or even the correct scaling. Furthermore, in real data the correct scaling may be masked because of noise and other data imperfections. It is recommended, if the data exhibit power-law spectra, to compare nonlinear prediction on the real data to nonlinear prediction on many fBm's which are as long as the real data and have similar power-law spectra. If the real data are not an fBm, significant differences will appear in the prediction curves. Other prediction approaches such as the local approximation produce similar conclusions.[214]

In summary, nonlinear prediction not only produces better forecasts, it also provides very strong evidence for or against the existence of low-dimensional chaotic attractors.

CHAPTER 11

OTHER DEVELOPMENTS AND TRENDS IN THE APPLICATION OF CHAOS

1. SHADOWING

Suppose we are given the mapping $x_{n+1} = f(x_n)$, which for some initial condition x_0 corresponds to a purely deterministic orbit or sequence $\{x_n\}$: $x_1, x_2, \ldots, x_n$. Due to roundoff and/or truncation errors, however, the sequence that we might observe when this mapping is iterated will be $\{y_n\}$: $y_1, y_2, \ldots, y_n$ (see Fig. 131).

Anosov[7] and Bowen[23] have shown that dynamical systems that are uniformly hyperbolic will have the *shadowing* property: if numerical errors are not large, each simulated trajectory will remain close to the noisy trajectory for ever. The problem is that, in reality, most chaotic systems are not uniformly hyperbolic and shadowing fails after some time. Thus, one should expect that after some time the simulated trajectories will not correlate with the "actual" trajectories. This time, however, can be very long.[98,99] For the Hénon map it is about 10^8 time steps.

The shadowing problem is that of finding the orbit $\{x_n\}$ that stays close to the orbit $\{y_n\}$. One way to do this is as follows: Define the Euclidean distance between two points y_n and x_n as $\|y_n - x_n\|$. Then we can define the distance between two trajectory segments as

$$D(x, y) = \sqrt{\frac{1}{n} \sum_{i=1}^{n} \|y_i - x_i\|^2}$$

To find an orbit x that "shadows" orbit y, we aim to find a segment x that

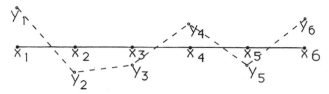

FIGURE 131. Due to measurement or roundoff or truncation error an actual orbit $\{x_n\}$ is not observed. Rather we observe a sequence $\{y_n\}$. The shadowing problem is that of finding the orbit $\{x_n\}$ that stays close to the orbit $\{y_n\}$.

minimizes $D(x, y)$ or $D^2(x, y)$ subject to the constraint that there exists a deterministic mapping of the form

$$x_{n+1} = f(x_n)$$

Solving a minimization problem with a constraint can be achieved with the help of the Lagrange multipliers (see Strang[199]). The procedure calls for minimizing the function L that takes into account the function to be minimized and the constraint

$$L = \text{the function to be minimized} - \lambda[\text{constraint}]$$

In our case, the problem is equivalent to minimizing

$$L = \sum_{i=1}^{n} \| y_i - x_i \|^2 + 2 \sum_{i=1}^{n-1} [f(x_i) - x_{i+1}]^T \lambda_i \qquad (11.1)$$

where λ_i's are Lagrange multipliers.[52] The superscript T denotes transpose. Differentiating Eq. (11.1) with respect to x_i and λ_i, we obtain

$$\frac{\partial L}{\partial x_i} = -(y_i - x_i) + f'^T(x_i)\lambda_i - \lambda_{i-1}$$

$$\frac{\partial L}{\partial \lambda_i} = f(x_i) - x_{i+1} \qquad (11.2)$$

where $f'(x_i) = (df/dx)_{(x_i)}$ is the Jacobian matrix of f at x_i. Setting Eqs. (11.2) to zero (i.e., asking for the extremum), we get

$$-(y_i - x_i) + f'^{\mathrm{T}}(x_i)\lambda_i - \lambda_{i-1} = 0$$

$$f(x_i) - x_{i+1} = 0 \qquad (11.3)$$

where in the first equation $i = 1, \ldots, n$ and in the second $i = 1, \ldots, n - 1$. Numerical solution of the nonlinear system of Eqs. (11.3) provides the shadowing trajectory $\{x_n\}$. One of the many ways to solve this system is given in Farmer and Sidorowich[56] and Eubank and Farmer.[52] According to their method, the above system is linearized by expanding $f(x_i)$ in a Taylor's series about a trial solution $\hat{x}_i$. After linearization, Eqs. (11.3) reduce to a matrix equation of the form $AB = C$ in which B contains an approximation to the true shadowing trajectory $\{x_n\}$.

2. NOISE REDUCTION

Noise is present in all physical systems. Noise is also present in numerical experiments as a result of roundoff and/or truncation errors. Therefore, we usually observe a noisy orbit

$$y_n = x_n + \epsilon_n$$

where x_n is the "clean" part from some deterministic system $x_{n+1} = f(x_n)$ and ϵ_n is the added noise.

The problem of noise reduction is to eliminate noise and retain an orbit that is as clean as possible. One common approach is to use Fourier analysis. Consider noise as a collection of high-frequency components and simply subtract them from the power spectra (or Fourier transform) of the components. Then invert the spectra to obtain a new time series with some of the high-frequency components removed. This might work in some instances, but we cannot forget that spectra from a low-dimensional chaotic attractor are often indistinguishable from those of random processes. Thus, suppression of high frequencies can modify the underlying dynamics. In fact, Badii et al.[12] demonstrated that such an approach might introduce an extra Lyapunov exponent that depends on the cutoff frequency.

Consequently, the noise reduction problem becomes a little more involved. One possible way is to reduce noise via shadowing. As we saw, a trajectory $\{x_n\}$ can be derived from a trajectory $\{y_n\}$ by solving the shadowing problem. Strictly speaking, producing a shadowing trajectory is not a noise reduction problem since the shadowing orbit $\{x_n\}$ is simply an artificial construction. We may, however, assume that $\{x_n\}$ is a signal and write

$$y_n = x_n + \xi_n$$

where $\xi_n = y_n - x_n$ is the "effective" noise. Then the procedure outlined in the previous section produces a trajectory close to the true trajectory (see also Farmer and Sidorowich[56]).

Another approach that "uses" shadowing for noise reduction is that of Hammel.[97] Hammel calls his algorithm "noise reduction by shadowing," and the idea behind this approach is as follows. Given a noisy 2D orbit $\{p_n\}$, $n = 0, \ldots, N$, a less noisy orbit $\{x_n\}$ is sought, where a noise-free orbit $\{x_n\}$ would satisfy $x_n = f(x_{n-1})$. Setting $\phi_n = x_n - p_n$ and $\Pi_n = f(p_{n-1}) - p_n$, it follows that

$$\phi_n = f(x_{n-1}) - f(p_{n-1}) + \Pi_n \tag{11.4}$$

Assuming that x_n and p_n are close to each other, we denote by L_n the linearized map at p_n and make the approximation

$$L_n \phi_n = f(x_n) - f(p_n) \tag{11.5}$$

Combining Eqs. (11.5) and (11.4), we obtain

$$\phi_n = L_{n-1}\phi_{n-1} + \Pi_n \tag{11.6}$$

Equation (11.6) is an iteration scheme for the unknown set $\{\phi_n\}$. Consequently, ϕ_n and Π_n are expressed via the formulas

$$\phi_n = a_n \hat{e}_n + b_n \hat{c}_n$$
$$\Pi_n = a'_n \hat{e}_n + b'_n \hat{c}_n \tag{11.7}$$

where $\{\hat{e}_n\}$ and $\{\hat{c}_n\}$ are unit vectors along the expanding and contracting directions. The orbit is two dimensional, and since it is chaotic there is expansion in one direction and contraction along the other. From Eqs. (11.6) and (11.7) we now have

$$\phi_{n+1} = L_n(a_n \hat{e}_n + b_n \hat{c}_n) + (a'_{n+1} \hat{e}_{n+1} + b'_{n+1} \hat{c}_{n+1})$$

Since the sequence $\{\Pi_n\}$ is known, $\{a'_n\}$ and $\{b'_n\}$ are also known. The sequences $\{a_n\}$ and $\{b_n\}$ can be determined recursively by using

$$a_{n+1} = \| L_n \hat{e}_n \| a_n + a'_{n+1}$$

$$b_{n+1} = \| L_n \hat{c}_n \| b_n + b'_{n+1}$$

Finally, those two equations can be cast into the following stable iterative scheme:

$$a_n = \frac{a_{n+1} - a'_{n+1}}{\| L_n \hat{e}_n \|}, \qquad a_N = 0$$

$$b_{n+1} = b_n \| L_n \hat{c}_n \| + b'_{n+1}, \qquad b_0 = 0$$

A different approach has been put forward by Farmer and Sidorowich.[54] Their philosophy is that a standard way to reduce noise is by averaging nearby points (for example, moving averaging). The problem is then reduced to finding the best way of averaging. What they propose is illustrated in Fig. 132. Row 1 shows three trajectory points at times $t - 1$, $t, t + 1$. Their observed values are x_{t-1}, x_t, x_{t+1}, respectively. These observed values can be decomposed into the "desired" deterministic part y (indicated by a solid circle) and the noise ξ (indicated by a surrounding cloud). Think of this cloud as representing some probability distribution of ξ at times $t - 1, t, t + 1$, etc. To produce a less noisy value at time t, we first take $x_{t-1} = y_{t-1} + \xi_{t-1}$ and apply to it the nonlinear mapping f, thus producing a new value x'_t. Since the mapping is nonlinear, by doing this we transform the probability distribution of the noise (the deterministic part should not change since the mapping is known). We may do the same to x_{t+1} by transporting it back to produce a value x''_t (row 3). Then average x'_t and x''_t to obtain a cleaner value x_t (row 4). Repeat this many times until a very clean orbit is obtained. Under the mapping f, a probability distribution has an induced transformation $\mathbf{f}$:

$$P(\xi_{t+1}) = P(x_{t+1} - y_{t+1}) = \mathbf{f}[P(x_t, y_t)] = \mathbf{f}[P(\xi_t)]$$

Following Farmer and Sidorowich,[54] suppose we transport mea-

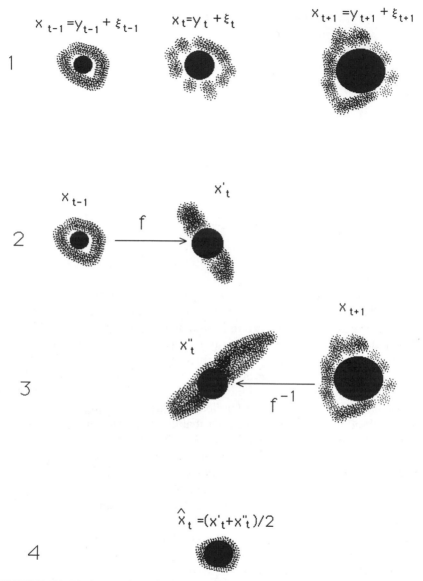

FIGURE 132. The idea behind reducing noise by nonlinear smoothing. The circles represent the true values, and the surrounding cloud the additive noise. As successive measurements are transported to the same point in time, the associated noise probability distribution changes. By averaging the values of the transported points, we may obtain a better estimate of the true value x_t.

surements from times $t - a, \ldots, t + b$, to time t. Since ξ_t are independent, the joint probability distribution is

$$\bar{P}(f^a(x_{t-a}), \ldots, f^{-b}(x_{t+b})) = A \prod_{j=a}^{-b} \mathbf{f}^j(P(x_{t-j} - y_{t-j})) \quad (11.8)$$

where A is some normalization constant and $a \geq 0$, $b \geq 0$. If n_t is small, we may linearize $\mathbf{f}$ to obtain

$$\mathbf{f}(P(x_t - y_t)) \approx A'P((f(x_t) - f(y_t)) \, df(x_t)^{-1})$$
$$= A'P((f(x_t) - y_{t+1}) \, df(x_t)^{-1}) \quad (11.9)$$

where A' is a new normalization factor and $df(x_t)$ is the derivative of f at x_t. Assuming that the noise ξ_t has a Gaussian distribution $N(0, \sigma^2)$, and taking into account Eqs. (11.8) and (11.9), we arrive at

$$\bar{P}(f^a(x_{t-a}), \ldots, f^{-b}(x_{t+b}))$$
$$\approx A \prod_{j=a}^{-b} \exp\left\{ \frac{-\|(f^i(x_{t-j}) - y_t)[df^j(x_{t-j})]^{-1}\|^2}{2\sigma^2} \right\}$$

The goal is to estimate y_t. Setting our estimate $Y_t - y_t$ and employing the maximum likelihood procedure, we ask that

$$\frac{\partial \log \bar{P}}{\partial Y_t} = 0$$

Solving this equation yields

$$Y_t = \left[\sum_{j=a}^{-b} \Theta_j \right]^{-1} \sum_{j=a}^{-b} \Theta_j f^j(x_{t-j})$$

where

$$\Theta = ([df^j(x_{t-j})]^T \, df^j(x_{t-j}))^{-1}$$

is a $d \times d$ symmetric matrix that depends on x_{t-j}.

All the foregoing approaches assume that the nonlinear mapping f is known. The results from applying these techniques are stunning. For examples see Figs. 133 and 134. The top figure shows the noisy orbit (obtained by adding noise to a clean orbit), and the bottom figure shows the orbit after noise reduction by the shadowing algorithm[97] is applied. Note how the noise reduction procedure preserves the fractal structure of the attractor.

Recently an approach to reduce noise without requiring that the underlying mapping be known has been proposed by Kostelich and Yorke.[119] Their approach is based on the same philosophy behind the local approximation for predicting time series. Consider an observable $s(t)$. We reconstruct the trajectory in an embedding space of $2D + 1$ dimensions, and for each reference point x we define a neighborhood of points. We then derive the linear mapping $x_{n+1} = f(x_n) = ax_n + b = L(x_n)$. The resulting

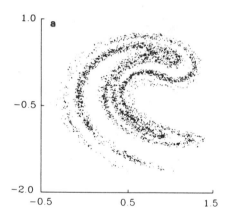

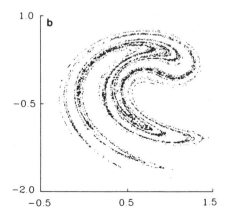

FIGURE 133. (a) A noisy orbit of N = 10,000 points for the Ikeda map. Normally distributed noise with standard deviation ≈ 0.4 was added to a computed orbit, corresponding to an initial noise level $\Delta \approx 0.051$. (b) The cleaned orbit after 24 iterations of the refinement process. The fractal structure of the attractor is apparent in the folded leaves of the unstable manifold. (Reproduced by permission from Hammel.[97])

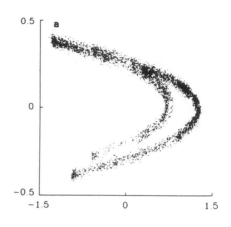

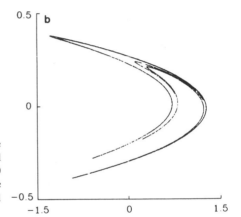

FIGURE 134. (a) A noisy orbit for the
Hénon map consisting of 10,000 points, and
$\Delta \approx 0.0455$. (b) The cleaned orbit after 20
refinements. The manifold structure of the
Hénon attractor is apparent. (Reproduced
by permission from Hammel.[97])

maps for the reference points along the trajectory are then stored, and the
trajectory at each point is perturbed to a value $\bar{x}$. The aim is to perturb
the trajectory in such a way so that it is more consistent with the dynamics.
This is achieved by minimizing the sum

$$\sum w \| \hat{x}_i - x_i \|^2 + \| x_i - L_{i-1}(\hat{x}_{i-1}) \| + \| \hat{x}_{i+1} - L_i(\hat{x}_i) \|^2$$

where w is a weighing factor and the sum runs over all points along the
trajectory. The equation can be solved by least squares (see Chapter 10).
The solution gives a new sequence $\{x_n\}$. Then the procedure is repeated
again and again until a much cleaner orbit is obtained. Figures 135 and
136 show results from this approach.

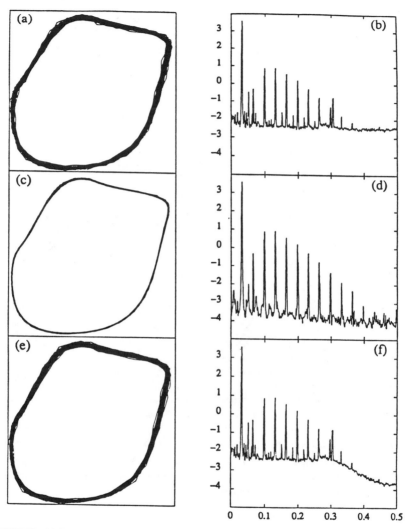

FIGURE 135. Phase portraits and power spectra for measurements of wavy vortex in a Couette-Taylor experiment. (a),(b) Phase portrait and power spectra before noise reduction is applied; (c),(d) after noise reduction; (e),(f) after a low-pass filter is applied to the original data. The vertical axis in (b),(d), and (f) is the base 10 logarithm of the power spectral density; the horizontal axis is in multiples of the Nyquist frequency. (Figure courtesy of Dr. E. J. Kostelich.)

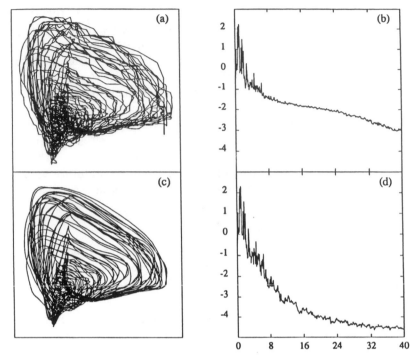

FIGURE 136. As in Fig. 135 but for a weakly chaotic flow in a Couette-Taylor experiment. (a),(b) before noise reduction; (c),(d) after noise reduction. (Figure courtesy of Dr. E. J. Kostelich.)

Figure 135a,b show the phase portrait and spectra for measurements of wavy (periodic) vortex flow in a Couette-Taylor experiment. Figure 135c,d show the orbit and spectra after the noise reduction algorithm of Kostelich and Yorke[119] is applied. Figure 135e, f show the orbit after the high-frequency components of the spectra in 135b are removed. Note that such a procedure does not result in a cleaner orbit. Figure 136 is similar, but is for measurements of chaotic flow in a Couette-Taylor experiment.

3. STATISTICAL NOISE REDUCTION

The aforementioned techniques remove noise from each and every point of a trajectory. In many cases, however, this may not be what we really want. Often we may only be interested simply in cleaner statistical properties of the time series, such as power spectra and dimension. For this purpose any cleaner time series with the same statistical properties as the "true" time series is good enough. This is called statistical noise reduction. A straightforward and simple approach for statistical noise reduction is to find a global model to the noisy data and iterate it (starting from some initial condition) to obtain a new time series. As demonstrated in Eubank and Farmer,[52] this type of noise reduction could be quite effective if the underlying attractor is not very sensitive to the parameters involved. Since statistical noise reduction produces a new orbit with the same statistical properties as the true orbit for every initial condition, it could be used in a way similar to the way bootstrapping is used to artificially obtain many data points. Such an approach might be useful for approximating invariant measures like dimensions from an initially small amount of data (see Casdagli[28]).

4. SOME ADDITIONAL COMMENTS ON THE EFFECT OF TRUNCATION AND ROUNDOFF ERROR ON SIMULATED CHAOTIC TRAJECTORIES

When a system is chaotic, a trajectory in its state-space (depicting the evolution of the system from some initial condition) never repeats or crosses itself. If that were to happen, the system would return to a state it was in at some time in its past, and then it would have to follow the same path. As a result, an aperiodic but deterministic evolution is obtained. The trajectory, however, may at some time in the future come very, very close to a state of the past. If those two states differ at the nth decimal point, then a computer that truncates or rounds off at the $(n-1)$st decimal point will, or may, make those two points equivalent.

This can be illustrated by considering the logistic equation $x_{n+1} = 4x_n(1 - x_n)$.[208] In this form the logistic equation is chaotic, i.e., aperiodic. Now assume that we desire to generate a long time series and that we are dealing with a computer that, in its calculations, carries only two decimal points. We then start with the initial condition $x_0 = 0.121134\cdots$. The computer reads this value as $x_0 = 0.12$ and produces a value of $x_1 = 0.42$ (the exact value is 0.4224, but it is truncated to 0.42). If we continue this iteration process, we find that $x_2 = 0.97$, $x_3 = 0.11$, $x_4 = 0.39$, $x_5 = 0.95$, $x_6 = 0.19$, $x_7 = 0.61$, $x_8 = 0.95$, $x_9 = 0.19$, $x_{10} = 0.61$, $x_{11} = 0.95$, and so on. The evolution has become periodic of period 3 after eight time steps. We can repeat this experiment many times (each time starting with a different initial condition) and find the mean number of time steps before the evolution becomes periodic. We then can proceed with other truncation scenarios and obtain a graph like Fig. 137, which shows $\log N$, the logarithm of the number of time steps before the chaotic evolution becomes periodic, as a function of n, the number of digits after the decimal pionts that are considered in the calculations.

As can be seen from Fig. 137, the evolution sooner or later becomes periodic. Roughly, one can consider N to vary as $N \propto 10^{n/2}$. Thus, with a computer that truncates at the 16th decimal point, a chaotic evolution becomes periodic after 10^8 time steps.

Similar results are obtained when we consider roundoff instead of truncation. Starting again with the initial condition $x_0 = 0.121134\cdots$ and assuming that our computer rounds off at the second decimal point, we obtain $x_1 = 0.42$, $x_2 = 0.97$, $x_3 = 0.12$, $x_4 = 0.42$, $x_5 = 0.97$, and so on. The evolution has again become periodic of period 3 after (in this case) four time steps. In general, it is found that roundoff-"induced" periodicity is achieved somewhat faster than truncation-"induced" periodicity. Take, for example, two states represented by the values $0.1212436\cdots$ and $0.1211733\cdots$ Rounding off at the fourth decimal point makes both

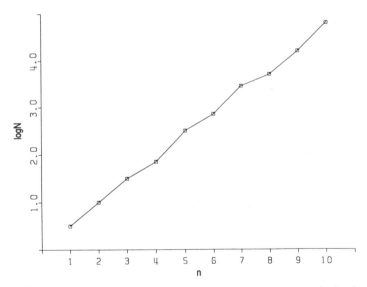

FIGURE 137. Truncation will cause any chaotic sequence to become periodic after some steps. In this figure the logarithm of the average number of steps ($\log N$) before a chaotic sequence from the logistic equation becomes periodic is plotted against the number of digits after the decimal point that are carried in the calculations (n) (Tsonis[208]).

these values equal to 0.1212, while truncation makes them equal to 0.1212 and 0.1211, respectively.

These results provide confidence in computer-generated chaotic trajectories for low-dimensional systems, since in such cases we do not usually require more than 10^8 data points (if a computer that carries 16 digits in its calculations is used). However, what causes shadowing to fail will also cause a chaotic trajectory to repeat (become periodic).

5. CONTROL OF CHAOS

In many instances the behavior of a dynamical system can change from periodic to chaotic (and vice versa), even with a very small change in the controlling parameter(s).

As discussed at the end of Chapter 6, a chaotic attractor has an infinite number of unstable periodic orbits embedded in it. The idea behind controlling chaos is to exploit these existing periodic orbits by perturbing some parameter of the system so as to stabilize one of these orbits. Then the system becomes periodic. Such a change in the dynamical behavior of the system might be desirable, since once the system becomes periodic it also becomes more predictable.

While clearly important, the problem of controlling chaos has only lately begun receiving attention. A theoretical treatment of the problem has recently been given by Ott et al.,[159] where the behavior of the Hénon map was controlled numerically. Experimental control of chaos was achieved by Ditto et al.[42] soon after. Their experimental system consisted of a gravitationally bucked, amorphous, magnetoelastic ribbon. The ribbon was placed within three mutually orthogonal pairs of Helmholtz coils, which allowed Ditto and his colleagues to compensate for the earth's magnetic field and to apply a uniform vertical magnetic field along the ribbon, causing the ribbon to oscillate. A sensor measured the curvature of the ribbon near its base. For some value of the magnetic field the oscillations are chaotic. From this output the period 1 and period 2 orbits were located, and the magnetic field was modified slightly to obtain a change first to period 1 oscillations and then to period 2 oscillations.

Other notable studies which differ in their approach that deal with the control of chaos include those by Hubler[105] and Fowler.[62]

REFERENCES

1. H. D. I. ABARBANEL, R. BROWN AND J. B. KADTKE, "Predictions and System Identification in Chaotic Nonlinear Systems: Time Series with Broadband Spectra," *Phys. Lett.* **A138**, 401–408 (1989).
2. H. D. I. ABARBANEL, R. BROWN AND J. B. KADTKE, "Prediction in Chaotic Systems: Methods for Time Series with Broadband Fourier Spectra," *Phys. Rev.* **A41**, 1782–1807 (1990).
3. H. D. I. ABARBANEL AND M. B. KENNEL, "Lyapunov Exponents in Chaotic Systems: Their Importance and Their Evaluation Using Observed Data," *Mod. Phys. Lett. B.,* **5**, 1347–1375 (1991).
4. R. H. ABRAHAM AND C. D. SHAW, *Dynamics: The Geometry of Behavior,* Parts I–IV (Aerial Press, Santa Cruz, California, 1988).
5. G. AHLERS AND R. BEHRINGER, "Evolution of Turbulence from the Rayleigh-Bénard Instability," *Phys. Rev. Lett.* **40**, 712–715 (1978).
6. G. AHLERS AND R. W. WALDEN, "Turbulence near the Onset of Convection," *Phys. Rev. Lett.* **44**, 445–449 (1980).
7. D. V. ANOSOV, "Geodesic Flows and Closed Riemannian Manifolds with Negative Curvature," *Proc. Inst. Math* **90**, (1967).
8. D. G. ARONSON, M. A. CHORY, G. R. HALL AND R. P. MCGHEE, Bifurcations from an Invariant Circle for Two-Parameter Families of Maps of the Plane: A Computer Assisted Study," *Comm. Math. Phys.* **83**, 303–311 (1982).
9. F. T. ARRECCHI, R. MEUCCI, G. PUCCIONI AND J. TREDDICE, "Experimental Evidence of Subharmonic Bifurcations, Multistability and Turbulence in a Q-Switched Gas Laser," *Phys. Rev. Lett.* **49**, 1217–1220 (1982).
10. D. AUERBACH, P. CVITANOVIC, J.-P. ECKMAN, G. H. GUNARATNE AND I. PROCACCIA, "Exploring Chaotic Motion through Periodic Orbits," *Phys. Rev. Lett.* **58**, 2387–2389 (1987).
11. A. BABLOYANTZ AND A. DESTEXHE, "Low-Dimensional Chaos in an Instance of Epilepsy," *Proc. Nat. Acad. Sci. USA* **83**, 3513–3517 (1986).
12. R. BADII, G. BROGGI, D. DERIGHETTI, M. RAVANI, S. CILIBERTO AND A. POLITI, "Dimension Increase in Filtered Chaotic Signals," *Phys. Rev. Lett.* **60**, 979–983 (1988).
13. R. BADII AND A. POLITI, "Statistical Description of Chaotic Attractors: The Dimension Function," *J. Stat. Phys.* **40**, 725–750 (1985).

14. R. P. BEHRINGER, S. D. MEYERS AND H. L. SWINNEY, "Chaos and Mixing in a Geo-
    strophic Flow," *Phys. Fluids* **A3**, 1243–1249 (1991).

15. E. BELTRAMI, *Mathematics for Dynamic Modeling* (Academic Press, Orlando, 1987).

16. G. BENETTIN, D. CASATI, L. GALGANI, A. GIORGILLI AND L. SIRONI, "Apparent Fractal
    Dimension in Conservative Dynamical Systems," *Phys. Lett.* **118A**, 325–330 (1986).

17. G. BENETTIN, L. GALGANI, A. GIORGILLI AND J. STRELCYN, "Lyapunov Characteristic
    Exponents for Smooth Dynamical Systems and for Hamiltonian Systems: A Method
    for Computing All of Them," *Meccanica* **15**, 9–20 (1980).

18. T. B. BENJAMIN AND T. MULLIN, "Anomalous Modes in the Taylor Experiment," *Proc.
    R. Soc. London* **A377**, 221–222 (1981).

19. P. BERGÉ, M. DUBOIS, P. MANNEVILLE AND Y. POMEAU, "Intermittency in Rayleigh-
    Bénard Convection," *J. Phys. Lett.* **41**, 341–345 (1980).

20. A. L. BERGER, "Obliquity and Precession for the Last 5,000,000 Years," *Astronomy
    Astrophys.* **51**, 1727–1735 (1976).

21. A. R. BISHOP, M. G. FOREST, D. W. MCLAUGHLIN AND E. A. OVERMAN, II, "A Quasi-
    Periodic Route to Chaos in Near-Integrable PDE," *Phys. D* **23**, 293–328 (1986).

22. A. R. BISHOP AND P. S. LOMDAHL, "Nonlinear Dynamics in Damped Sine-Gordon
    Systems," *Phys. D* **18**, 54–66 (1986).

23. R. BOWEN, "ω-Limit Sets for Axiom A Diffeomorphisms," *J. Diff. Eqs.* **18**, 333–339
    (1975).

24. W. A. BROCK, "Distinguishing Random and Deterministic Systems," *J. Econom. Theory*
    40, 168–195 (1986).

25. W. A. BROCK AND C. L. SEYERS, "Is the Business Cycle Characterized by Deterministic
    Chaos? Technical Paper 8617, Social Systems Research Institute, University of Wisconsin-
    Madison, Madison, Wisconsin, 1986.

26. D. S. BROOMHEAD AND G. P. KING, "Extracting Qualitative Dynamics from Experi-
    mental Data," *Phys. D* **20**, 217–226 (1986).

27. R. BROWN, P. BRYANT AND H. D. I. ABARBANEL, "Computing the Lyapunov Spectrum
    of a Dynamical System from Observed Time Series," *Phys. Rev.* **43**, 27–87 (1991).

28. M. CASDAGLI, "Nonlinear Prediction of Chaotic Time Series," *Phys. D* **35**, 335–356
    (1989).

29. M. CASDAGLI, "A Dynamical Systems Approach to Modeling Driven Systems," in *Non-
    linear Prediction and Modelling,* M. Casdagli and S. Eubank, Eds. (Addison-Wesley,
    Reading Massachusetts, 1991).

30. M. CASDAGLI, D. DES JARDIN, S. EUBANK, J. D. FARMER, J. GIBSON, N. HUNTER AND
    J. THEILER "Nonlinear Modeling of Chaotic Time Series: Theory and Applications, Los
    Alamos National Laboratory Preprint, LA-UR-91-1637, 1991.

31. M. CASDAGLI, J. GIBSON, D. FARMER AND N. HUNTER "State Space Reconstruction
    in the Presence of Noise," Los Alamos National Laboratory, reprint, LA-UR-91-1010,
    1990.

32. H. CHATÉ AND P. MANNEVILLE, "Transition to Turbulence via Spatiotemporal Inter-
    mittency," *Phys. Rev. Lett.* **58**, 112–115 (1986).

33. A. A. CHERNIKOV, R. Z. SAGDEEV, D. A. USIKOV AND G. M. ZAVLASKY, *Sov. Sci. Rev.
    C. Math. Phys.* **8**, 83–172 (1989).

34. B. V. CHIRIKOV, "A Universal Instability of Many Dimensional Oscillator Systems,"
    Phys. Rep. **52**, 263–379 (1979).

35. S. CILIBERTO AND M. A. RUBIO "Local Oscillations, Traveling Waves, and Chaos in
    Rayleigh–Bénard Convection," *Phys. Rev. Lett.* **58**, 2652–2655 (1987).

36. E. A. CODDINGTON AND N. LEVINSON, *Theory of Ordinary Differential Equations* (McGraw-Hill, New York, 1955).

36a.A. L. GOLDBERGER, D. R. RIGNEY, AND B. J. WEST, "Chaos and Fractals in Human Physiology," *Sci. Am.* **262**(2), 43–49 (1990).

37. J. P. CRUTCHFIELD, D. J. FARMER, N. H. PACKARD AND R. S. SHAW, "Chaos," *Sci. Am.* **254**(12), 46–57 (1986).

38. A. CUMMING AND P. S. LINSAY, "Quasiperiodicity and Chaos in a System with Three Competing Frequencies," *Phys. Rev. Lett.* **60,** 2719–2722 (1988).

39. P. CVITANOVIC, G. H. GUNARATNE AND I. PROCACCIA, "Topological and Metric Properties of Hénon-Type Strange Attractors," *Phys. Rev. A* **38**, 1503–1520 (1988).

40. D. DANGOISSE, P. GLORIEUX AND D. HANNEQUIN, "Laser Chaotic Attractors in Crisis," *Phys. Rev. Lett.* **57**, 2657–2660 (1986).

41. D. DEL-CASTILLO-NEGRETE AND P. J. MORRISON, "Hamiltonian Chaos and Transport in Quasigeostrophic Flows," *Proceedings of the American Institute of Physics,* La Jolla, California, 1991.

42. W. L. DITTO, S. N. RAUSEO AND M. L. SPANO, "Experimental Control of Chaos," *Phys. Rev. Lett.* **65,** 3211–3214 (1990).

43. R. J. DONNELLY, K. PARK, S. SHAW AND R. W. WALDEN, "Early Nonperiodic Transitions in Couette Flow," *Phys. Rev. Lett.* **44**, 987–991 (1980).

44. M. DUBOIS, M. A. RUBIO AND P. BERGÉ, "Experimental Evidence of Intermittencies Associated with a Subharmonic Bifurcation," *Phys. Rev. Lett.* **51**, 1446–1449 (1983).

45. J.-P. ECKMAN, S. O. KAMPHORST, D. RUELLE AND S. CILILBERTO, "Liapunov Exponent from Time Series," *Phys. Rev. A* **34**, 4971–4980 (1986).

46. J.-P. ECKMAN AND D. RUELLE, "Ergodic Theory of Chaos and Strange Attractors," *Rev. Mod. Phys.* **57**, 617–628 (1985).

47. J. B. ELSNER AND A. A. TSONIS, "Nonlinear Prediction, Chaos and Noise," *Bull. Am. Met. Soc.,* **73,** 49–60 (1992).

48. I. R. EPSTEIN "Oscillations and Chaos in Chemical Systems," *Phys. D* **7**, 47–51 (1983).

49. C. ESSEX, T. LOOKMAN AND M. A. H. NERENBERG, "The Climate Attractor over Short Time Scales," *Nature* **326**, 64–66 (1987).

50. C. ESSEX AND M. A. H. NERENBERG, "Fractal Dimension: Limit Capacity or Hausdorff-Dimension?" *Am. J. Phys.* **58**, 986–988 (1990).

51. C. ESSEX AND M. A. H. NERENBERG, "Comments on Deterministic Chaos: The Science and the Fiction by D. Ruelle," *Proc. R. Soc. London* **A435**, 287–292 (1991).

52. S. EUBANK AND J. D. FARMER, "An Introduction to Chaos and Randomness," Los Alamos National Laboratory, reprint LA-UR-90-1874, 1990.

53. J. D. FARMER AND J. J. SIDOROWICH, "Predicting Chaotic Time Series," *Phys. Rev. Lett.* **59**, 845–848 (1987).

54. J. D. FARMER AND J. J. SIDOROWICH, "Exploiting Chaos to Predict the Future and Reduce Noise," Los Alamos National Laboratory, scientific report, LA-UR-88-901, 1988a.

55. J. D. FARMER AND J. J. SIDOROWICH, "Exploiting Chaos to Predict the Future and Reduce Noise" In *Evolution, Learning and Cognition,* Y. C. Lee, Ed. (World Scientific Press, New York, 1988).

56. J. D. FARMER AND J. J. SIDOROWICH, "Optimal Shadowing and Noise Reduction," Los Alamos National Laboratory. preprint, LA-UR-90-053, 1990.

57. J. FEDER, *Fractals* (Plenum, New York, 1988).

58. M. J. FEIGENBAUM, "Universal Behavior in Nonlinear systems," *Los Alamos Sci.* **1,** 4–27 (1980).

59. M. J. FEIGENBAUM, L. P. KADANOFF AND S. J. SHENKER, "Quasiperiodicity in Dissipative Systems," *Phys. D* **5**, 370–381 (1983).

60. P. R. FENSTERMACHER, H. L. SWINNEY AND J. P. GOLLUB, "Dynamical Instabilities and the Transition to Chaotic Taylor Vortex Flow," *J. Fluid Mech.* **94**, 103–117 (1979).

61. J. FORD "Chaos: Solving the Unsolvable Predicting the Unpredictable," *Chaotic Dynamics and Fractals,* M. F. Barnsley and S. G. Demko, Eds. (Academic Press, San Diego, California, 1986).

62. T. B. FOWLER, "Application of stochestic control techniques to chaotic nonlinear systems," *IEEE Trans. Autom. Control.* **34**, 201–210 (1989).

63. K. FRAEDRICH, "Estimating the Dimensions of Weather and Climate Attractors," *J. Atmos. Sci.* **43**, 419–432 (1986).

64. K. FRAEDRICH, "Estimating Weather and Climate Predictability on Attractors," *J. Atmos. Sci.* **44**, 722–728 (1987).

65. G. W. FRANK, "Recovering the Lyapunov Exponent from Chaotic Time Series," Ph.D. thesis, University of Western Ontario, London, Ontario, 1990.

66. G. W. FRANK, T. LOOKMAN, M. A. H. NERENBERG AND C. ESSEX, "Chaotic Time Series Analysis of Epileptic Seizures," *Phys. D* **46**, 427–438 (1990).

67. A. M. FRAZER "Reconstructing Attractors from Scalar Time Series: A Comparison of Singular System and Redundancy Criteria," *Phys. D* **34**, 391–404 (1989).

68. A. M. FRAZER AND H. SWINNEY, "Independent Coordinates for Strange Attractors from Mutual Information," *Phys. Rev. A* **33**, 1134–1140 (1986).

69. P. FREDERICKSON, J. KAPLAN, E. YORKE AND J. YORKE, "The Lyaponov Dimension of Strange Attractors," *J. Diff. Eqs.* **49**, 185–192 (1983).

70. W. J. FREEMAN, "The Physiology of Perception," *Sci. Am.* **264**(2), 78–85 (1991).

71. U. FRISCH, Z. S. SHE AND O. THUAL, "Viscoelastic Behavior of Cellular Solutions to the Kuramoto-Sivanshinsky Model," *J. Fluid Mech.* **168**, 221–240 (1986).

72. T. FRISON, "Predicting Nonlinear and Chaotic Systems Behavior Using Neural Networks," *J. Neural Net. Comput.* **2**, 31–39 (1990).

73. H. FROEHLING, J. P. CRUTCHFIELD, J. D. FARMER, N. H. PACKARD AND R. S. SHAW, "On Determining the Dimension of Chaotic Flows," *Phys. D* **3**, 605–611 (1981).

74. M. GIGLIO, S. MUSAZZI AND V. PERINI, "Transition to Chaotic Behavior via a Reproducible Sequence of Period-Doubling Bifurcations," *Phys. Rev. Lett.* **47**, 243–246 (1981).

75. L. GLASS, M. R. GUEVARA AND A. SHRIER, "Bifurcation and Chaos in a Periodically Stimulated Cardiac Oscillator," *Phys. D* **7**, 89–97 (1983).

76. A. L. GOLDBERGER, D. R. RIGNEY AND B. J. WEST, "Chaos and Fractals in Human Physiology," *Sci. Am.* **263**(2), 43–49 (1990).

77. J. P. GOLLUB, "What Causes Noise in a Convecting Fluid," *Phys. A* **118**, 28–37 (1983).

78. J. P. GOLLUB AND S. V. BENSON, "Many Routes to Turbulent Convection," *J. Fluid Mech.* **100**, 449–470 (1980).

79. J. P. GOLLUB, S. V. BENSON AND J. F. STEINMAN, "A Subharmonic Route to Turbulent Convection," *Ann. N.Y. Acad. Sci.* **357**, 22–32 (1980).

80. J. P. GOLLUB, A. R. MCCARRIAR AND J. F. STEINMAN, "Convective Pattern Evolution and Secondary Instabilities," *J. Fluid Mech.* **125**, 259–267 (1982).

81. J. P. GOLLUB, E. J. ROMER AND J. E. SOCOLAR, "Trajectory Divergence for Coupled Relaxation Oscillators: Measurements and Models," *J. Stat. Phys.* **23**, 321–333 (1980).

82. J. P. GOLLUB AND H. L. SWINNEY, "Onset of Turbulence in a Rotating Fluid," *Phys. Rev. Lett.* **35**, 927–930 (1975).

83. M. GORMAN, L. A. REITH AND H. L. SWINNEY, "Modulation Patterns, Multiple Fre-

quencies and Other Phenomena in Circular Couette Flow," *Ann. N.Y. Acad. Sci.* **357,** 10–21 (1980).

84. P. GRASSBERGER, "Generalized Dimensions of Strange Attractors," *Phys. Lett. A.* **97,** 227–230 (1983).

85. P. GRASSBERGER, "Do Climatic Attractors Exist?" *Nature* **323,** 609–612 (1986).

86. P. GRASSBERGER, "An Optimized Box-Assisted Algorithm for Fractal Dimensions," *Phys. Lett. A* **148,** 63–68 (1990).

87. P. GRASSBERGER AND I. PROCACCIA, "Measuring the Strangeness of Strange Attractors," *Phys. D* **9,** 189–208 (1983).

88. P. GRASSBERGER AND I. PROCACCIA, "Characterization of Strange Attractors," *Phys. Rev. Lett.* **50,** 346–349 (1983).

89. C. GREBOGI, S. W. MCDONALD, E. OTT AND J. A. YORKE, "Final State Sensitivity: An Obstruction to Predictability," *Phys. Lett. A* **99,** 415–419 (1983).

90. C. GREBOGI, E. OTT AND J. A. YORKE, "Fractal Basin Boundaries, Long-Lived Chaotic Transients and Unstable-Unstable Pair Bifurcation," *Phys. Rev. Lett.* **50,** 935–939 (1983).

91. C. GREBOGI, E. OTT AND J. A. YORKE, "Are Three Frequency Quasiperiodic Orbits to Be Expected in Typical Dynamical Systems? *Phys. Rev. Lett.* **51,** 339–342 (1983).

92. C. GREBOGI, E. OTT AND J. A. YORKE, "Chaos, Strange Attractors and Fractal Basin Boundaries in Nonlinear Dynamics," *Science* **238,** 632–638 (1987).

93. H. S. GREENSIDE, A. WOLF, J. SWIFT AND T. PIGNARATO, "The Impracticability of a Box-Counting Algorithm for Calculating the Dimensionality of Strange Attractors. *Phys. Rev. A* **25,** 3453–3456 (1982).

94. M. R. GUEVARA, L. GLASS AND A. SHRIER, "Phase Locking, Period-Doubling Bifurcations and Irregular Dynamics in Periodically Stimulated Cardiac Cells. *Science* **214** 1350–1353 (1981).

95. G. H. GUNARATNE AND I. PROCACCIA, "Organization of Chaos," *Phys. Rev. Lett.* **59,** 1377–1380 (1987).

96. E. G. GWINN AND R. M. WESTERVELT, "Intermittent Chaos and Low-Frequency Noise in the Driven Damped Pendulum," *Phys. Rev. Lett.* **54,** 1613–1617 (1985).

97. S. M. HAMMEL, "A Noise Reduction Method for Chaotic Systems," *Phys. Lett. A* **148,** 421–428 (1990).

98. S. M. HAMMEL, J. A. YORKE AND C. GREBOGI, "Do Numerical Orbits of Chaotic Dynamical Processes Represent True Orbits?" *J. Complexity* **3,** 136–145 (1987).

99. S. M. HAMMEL, J. A. YORKE AND C. GREBOGI, "Numerical Orbits of Chaotic Processes Represent True Orbits," *Bull. Am. Math. Soc.* **19,** 465–469 (1988).

100. H. HAUCKE AND Y. MAENO, "Phase Space Analysis of Convection in ^{3}He-Superfluid ^{4}He Solution," *Phys. D* **7,** 69–72 (1983).

101. M. HÉNON, "A Two-Dimensional Mapping with a Strange Attractor," *Comm. Math. Phys.* **50,** 69–77 (1976).

102. H. G. E. HENTSCHEL AND I. PROCACCIA, "The Infinite Number of Generalized Dimensions of Fractals and Strange Attractors," *Phys. D* **8,** 435–444 (1983).

103. F. HESLOT, B. CASTAING AND A. LIBCHABER, "Transitions to Turbulence in Helium Gas," *Phys. Rev. A* **36,** 5870–5873 (1987).

104. E. HOPF, "Abzweigung einer periodischen Lösung von einer stationären Lösung eines differentialsystems," *Ber. Math.-Phys. Klasse Sachs. Akad. Wiss. Leipzig* **94,** 1–22 (1942). English translation in Ref. 143.

105. A. HÜBLER, "Adaptive Control of Chaotic Systems," *Helv. Phys. Acta* **62,** 343–346 (1989).

106. J. L. HUDSON, M. HART AND D. MARINKO, "An Experimental Study of Multiple Peak Periodic and Nonperiodic Oscillations in the Belousov-Zhabotinski Reaction," *J. Chem. Phys.* **71**, 1601–1610 (1979).

107. J. L. HUDSON AND J. C. MANKIN, "Chaos in the Belousov-Zhabotinsky Reaction," *J. Chem. Phys.* **74**, 6171–6177 (1981).

108. J. L. HUDSON AND O. E. RÖSSLER, "Chaos and Complex Oscillations in Stirred Chemical Reactors," in *Dynamics of Nonlinear Systems,* V. Hlavacek, Ed. (Gordon and Breach, New York, 1985).

109. N. F. HUNTER, "Application of Nonlinear Time Series Models to Driven Systems," in *Nonlinear Prediction and Modeling,* M. Casdagli and S. Eubank, Eds. (Addison-Wesley, Reading, Massachusetts, 1991).

110. J. M. HYMAN AND B. NICOLAENKO, "The Kuramoto-Sivashinsky Equation: A Bridge between PDE's and Dynamical Systems," *Phys. D* **18**, 113–126 (1986).

111. M. IANSITI, Q. HU, R. M. WESTERVELT AND M. TINKHAM, "Noise and Chaos in a Fractal Basin Boundary Regime of a Josephson Junction," *Phys. Rev. Lett.* **55**, 746–749 (1985).

112. H. IKEZI, J. S. DEGRASSE AND T. H. JENSEN, "Observation of Multiple-Valued Attractors and Crises in a Driven Nonlinear Circuit," *Phys. Rev. A* **28**, 1207–1209 (1983).

113. C. JEFFRIES AND J. PEREZ, "Observation of a Pomeau-Manneville Intermittent Route to Chaos in a Nonlinear Oscillator," *Phys. Rev. A* **26**, 2117–2122 (1982).

114. C. JEFFRIES AND J. PEREZ, "Direct Observation of Crises of the Chaotic Attractor in a Nonlinear Oscillator," *Phys. Rev. A* **27**, 601–603 (1983).

115. R. W. KATZ, "Statistical Evaluation of Climate Experiments with General Circulation Models: A Parametric Time Series Modeling Approach," *J. Atmos. Sci.* **39**, 1445–1455 (1982).

116. J. D. KEELER, "A Dynamical Systems View of Cerebellar Function," *Phys. D* **42**, 396–410 (1990).

117. C. L. KEPPENNE AND C. NICOLIS, "Global Properties and Local Structure of the Weather Attractor over Western Europe," *J. Atmos. Sci.* **46**, 2356–2370 (1989).

118. G. KING AND H. L. SWINNEY, "Limits of Stability and Defects in Wavy Vortex Flow," *Phys. Rev. A* **27**, 1240–1252 (1983).

119. E. J. KOSTELICH AND J. A. YORKE, "Noise Reduction: Finding the Simplest Dynamical System Consistent with the Data," *Phys. D* **41**, 183–196 (1990).

120. Y. KURAMOTO, "Diffusion-Induced Chaos in Reaction Systems," *Progr. Theor. Phys. Suppl.* **64**, 346–307 (1978).

121. Y. KURAMOTO AND T. TZUZUKI, "Persistent Propagation of Concentration Waves in Dissipative Media Far from Thermal Equilibrium," *Prog. Theor. Phys.* **55**, 356–369 (1976).

122. D. P. LATHROP AND E. J. KOSTELICH, "Characterization of an Experimental Strange Attractor by Periodic Orbits," *Phys. Rev. A* **40**, 4028–4031 (1989).

123. W. LAUTERBORN AND E. CRAMER, "Subharmonic Route to Chaos Observed in Acoustics," *Phys. Rev. Lett.* **47**, 1445–1448 (1981).

124. F. LEDRAPPIER, "Some Relations between Dimension and Lyapunov Exponents," *Comm. Math. Phys.* **81**, 229–232 (1981).

125. A. LIBCHABER, "From Chaos to Turbulence in Bénard Convection," *Proc. R. Soc. London A* **43**, 63–69 (1987).

126. A. LIBCHABER, S. FAUVE AND C. LAROCHE, "Two-parameter Study of the Routes to Chaos," *Phys. D* **7**, 73–82 (1983).

127. A. LIBCHABER, C. LAROCHE AND S. FAUVE, Period-doubling Cascade in Mercury, a Quantitative Measurement," *J. Phys.-Lett.* **43**, 211–216 (1982).

128. A. LIBCHABER AND J. MAURER, "Local Probe in a Rayleigh-Bénard Experiment in Liquid Helium," *J. Phys.-Lett.* **39**, 369–373 (1978).

129. A. LIBCHABER AND J. MAURER, "One Experience de Rayleigh-Benard de Geometrie Reduite; Multiplication, accrochage, et demultiplication de frequencies," *J. Phys.* **41**, C3, 51, (1980).

130. A. LIBCHABER AND J. MAURER, "A Rayleigh Benard Experiment: Helium in a Small Box," in *Nonlinear Phenomena at Phase Transitions and Instabilities* (T. Riste, Ed.), Plenum, New York, 1982.

131. A. J. LICHTENBERG AND M. A. LIEBERMAN, *Regular and Stochastic Motion* (Springer-Verlag, New York, 1983).

132. W. LIEBERT AND H. G. SCHUSTER, "Proper Choice of the Time Delay for the Analysis of Chaotic Time Series," *Phys. Lett. A* **142**, 107–111 (1989).

133. L. S. LIEBOVITCH AND T. TOTH, "A Fast Algorithm to Determine Fractal Dimensions by Box Counting," *Phys. Lett. A* **141**, 386–390 (1989).

134. P. S. LINSAY, "Period Doubling and Chaotic Behavior in a Driven Anharmonic Oscillator," *Phys. Rev. Lett.* **47**, 1349–1352 (1981).

135. E. N. LORENZ, "Deterministic Nonperiodic Flow," *J. Atmos. Sci.* **20**, 130–141 (1963).

136. E. N. LORENZ, "Atmospheric Predictability as Revealed by Naturally Occurring Analogues," *J. Atmos. Sci.* **26**, 636–646 (1969).

137. E. N. LORENZ, "Noisy Periodicity and Reverse Bifurcation," *Ann. N.Y. Acad. Sci.* **357**, 282–291 (1980).

138. E. N. LORENZ, "Dimension of Weather and Climate Attractors," *Nature* **353**, 241–242 (1991).

139. A. LORENZEN, G. PFISTER AND T. MULLIN, "End Effects on the Transition Time-Dependent Motion in the Taylor Experiment," *Phys. Fluids* **26**, 10–21 (1983).

140. B. B. MANDELBROT, *Fractals, Form, Chance, and Dimension* (W. H. Freeman, San Francisco, 1977).

141. B. B. MANDELBROT, *The Fractal Geometry of Nature* (W. H. Freeman, San Francisco, 1983).

142. B. B. MANDELBROT, "Fractals in Physics: Squig Clusters, Diffusions, Fractal Measures and the Unicity of Fractal Dimensionality," *J. Stat. Phys.* **34**, 859–930 (1984).

143. J. E. MARSDEN AND M. MCCRACKEN, *The Hopf Bifurcation and Its Applications* (Springer-Verlag, New York, 1976).

144. J. MAURER AND A. LIBCHABER, "Rayleigh-Bénard Experiment in Liquid Helium; Frequency Locking and the Onset of Turbulence," *J. Phys. Lett.* **40**, 419–423 (1979).

145. J. MAURER AND A. LIBCHABER, "Effect of the Prandtl Number on the Onset of Turbulence in Liquid Helium," *J. Phys. Lett.* **41**, 515–519.

146. R. M. MAY, "Simple Mathematical Models with Very Complicated Dynamics," *Nature* **261**, 459–467 (1976).

147. A. MEES, P. RAPP AND L. JENNINGS, "Singular-Value Decomposition and Embedding Dimension," *Phys. Rev. A* **36**, 340–352 (1987).

148. F. C. MOON, *Chaotic Vibrations* (Wiley, New York, 1987).

149. F. C. MOON AND G.-X. LI, "Fractal Basin Boundaries and Homoclinic Orbits for Periodic Motion in a Two-Well Potential," *Phys. Rev. Lett.* **55**, 1439–1442 (1985).

150. M. D. MUNDT, W. B. MAGUIRE, II, AND R. R. P. CHASE, "Chaos in the Sunspot Cycle: Analysis and Prediction," *J. Geophys. Res.* **96**, 1705–1716 (1991).

151. M. A. H. NERENBERG AND C. ESSEX, "Correlation Dimension and Systematic Geometric Effects," *Phys. Rev. A* **42,** 7065–7074 (1990).

152. J. M. NESE, "Quantifying Local Predictability in Phase Space," *Phys. D* **35,** 237–250 (1989).

153. S. NEWHOUSE, D. RUELLE AND F. TAKENS, "Occurrence of Strange Axiom A Attractors near Quasi-Periodic Flow on T^m, $m \geq 3$," *Comm. Math. Phys.* **64,** 449–454 (1978).

154. C. NICOLIS AND G. NICOLIS, "Is There a Climatic Attractor?" *Nature* **331,** 529–532 (1984).

155. A. R. OSBORNE AND R. CAPONIO, "Fractal Trajectories and Anomalous Diffusion for Chaotic Motions in 2D Turbulence," *Phys. Rev. Lett.* **64,** 1733–1736 (1990).

156. A. R. OSBORNE, A. D. KIRWAN, JR., A. PROVENZALE AND L. BERGAMASCO, "A Search for Chaotic Behavior in Large and Mesoscale Motions in the Pacific Ocean," *Phys. D* **23,** 75–83 (1986).

157. A. R. OSBORNE AND A. PROVENZALE, "Finite Correlation Dimension for Stochastic Systems with Power-law Spectra," *Phys. D* **35,** 357–381 (1989).

158. S. OSTLUND, D. RAND, J. SETHNA AND E. SIGGIA, "Universal Properties of the Transition from Quasiperiodicity to Chaos in Dissipative systems," *Phys. D* **8,** 303–309 (1983).

159. E. OTT., C. GREBOGI AND J. A. YORKE, "Controlling Chaos," *Phys. Rev. Lett.* **64,** 1196–1199 (1990).

160. J. M. OTTINO, "The Mixing of Fluids," *Sci. Am.* **259**(1), 56–67 (1989).

161. A. J. OWENS AND D. L. FILKIN, "Efficient Training of the Back-propagation Network by Solving a System of Stiff Ordinary Differential Equations," International Conference on Neural Networks, **2,** 381–386 (1989).

162. N. H. PACKARD, J. P. CRUTCHFIELD, J. D. FARMER AND R. S. SHAW, "Geometry from a Time Series," *Phys. Rev. Lett.* **45,** 712–716 (1980).

162a. G. P. PAULOS, G. A. KYRIAKOU, A. G. RIGAS, P. I. LIATSIS, P. C. TROCHOUTOS AND A. A. TSONIS, "Evidence for strange attractor structures in space plasma," *Ann. Geophys.* in press.

163. H. O. PEITGEN AND P. M. RICHTER, *The Beauty of Fractals* (Springer-Verlag, Berlin, 1986).

164. H. O. PEITGEN AND D. SAUPE, (Eds). *The Science of Fractal Images* (Springer-Verlag, New York, 1988).

165. J. C. PERRETT AND J. T. P. VAN STEKELENBORG, "A neural network as a Model for the Prediction of Sunspot Numbers," Bartol Research Foundation, University of Delaware, Newark, Delaware, 1990.

166. R. PFEFFER, G. BUZYNA AND R. KUNG, "Time-dependent Modes of Behavior of Thermally Driven Rotating Fluids," *J. Atmos. Sci.* **37,** 129–2149 (1980a).

167. R. PFEFFER, G. BUZYNA AND R. KUNG, "Relationships among Eddy Fluxes of Heat, Eddy Temperature Variances and Basic-state Temperature Parameters in Thermally Driven Rotating Fluids," *J. Atmos. Sci.* **37,** 2577–2599 (1980b).

168. C. A. PICKOVER, "A Note on Rendering 3-D Strange Attractors," *Comput. Graph.* **12,** 263–267 (1988).

169. Y. POMEAU AND P. MANNEVILLE, "Intermittent Transition to Turbulence in Dissipative Dynamical Systems," *Comm. Math. Phys.* **74,** 189–197 (1980).

169a. Y. POMEAUX, J. C. ROUX, A. ROSSI, S. BACHELART AND C. VIDAL, "Intermittent behavior in the Belousov–Zhabotinsky reaction," *J. Phys. Lett.* **42,** 271–273 (1981).

170. M. J. D. POWELL, "Radial Basis Function for Multivariate Interpolation: A Review," technical report, University of Cambridge.

171. W. H. PRESS, B. P. FLANNEY, S. A. TEUKOLSKY AND W. T. VETTERING, *Numerical Recipes* (Cambridge University Press, Cambridge, 1986).
172. P. E. RAPP, I. D. ZIMMERMAN, A. M. ALBANO, G. C. DEGUZMAN AND N. N. GREENBAUM, "Dynamics of Spontaneous Neural Activity in the Simian Motor Cortex: the Dimension of Chaotic Neurons," *Phys. Lett. A* **110**, 335–338 (1985).
173. D. A. ROBERTS, D. N. BAKER AND A. J. KLIMAS, "Indications of Low Dimensionality in Magnetospheric Dynamics," *Geophys. Res. Lett.* **18**, 151–154 (1991).
174. R. W. ROLLINS AND E. R. HUNT, "Intermittent Transient Chaos at Interior Crisis in the Diode Resonator," *Phys. Rev. A,* **29**, 3327–3334 (1984).
175. R. ROSEN, *Dynamical System Theory in Biology* (Wiley-Interscience, New York, 1970).
176. O. E. RÖSSLER, "An Equation for Continuous Chaos," *Phys. Lett. A* **57**, 397–398 (1976).
177. J. C. ROUX, "Experimental Studies of Bifurcations Leading to Chaos in the Belousov-Zhabotinsky Reaction," *Phys. D* **7**, 57–68 (1983).
178. J. C. ROUX, P. DEKEPPER AND H. L. SWINNEY, "Type II Intermittency in the Belousov-Zhabotinskii Reaction," *Phys. D* **7**, 57–68 (1983b).
179. J. C. ROUX, A. ROSSI, S. BACHELART AND C. VIDAL, "Representation from an Experimental Study of Chemical Turbulence," *Phys. Lett. A* **77**, 391–393 (1980).
180. J. C. ROUX, R. H. SIMOYI AND H. L. SWINNEY, "Observation of a Strange Attractor," *Phys. D* **8**, 257–262 (1983a).
181. J. C. ROUX, J. S. TURNER, W. D. MCCORMICK, AND H. L. SWINNEY, "Experimental Observations of Complex Dynamics in a Chemical Reaction," in *Nonlinear Problems: Present and Future,* A. R. Bishop, D. K. Campbell and B. Nicolaenko, Eds. (North-Holland, Amsterdam, 1982).
182. W. F. RUDDIMAN, M. RAYMO AND A. MCINTYRE, "Matuyama, 41000-Year Cycles: North Atlantic and Northern Hemisphere Ice Sheets," *Earth Planet. Sci. Lett.* **80**, 117–129 (1986).
183. D. RUELLE, "Chemical Kinetics and Differentiable Dynamical Systems," in *Nonlinear Phenomena in Chemical Dynamics,* A. Pacault and C. Vidal, Eds. (Springer-Verlag, Berlin, 1981).
184. D. RUELLE, *Chaotic Evolution and Strange Attractors* (Cambridge University Press, Cambridge, 1989).
185. D. RUELLE, "Deterministic Chaos: The Science and the Fiction," *Proc. R. Soc. London A* **427**, 241–248 (1990).
186. D. RUELLE AND F. TAKENS "On the Nature of Turbulence," *Comm. Math. Phys.* **20**, 167–172 (1971).
187. D. E. RUMELHART, G. E. HINTON AND R. J. WILLIAMS, "Learning Representations by Back-Progating Errors," *Nature* **323**, 533–536 (1986).
188. B. SALTZMAN, "Finite Amplitude Free Convection as an Initial Value Problem," *J. Atmos. Sci.* **19**, 329–341 (1962).
189. M. SANO AND W. SAWADA, "Measurement of the Lyapunov Spectrum from a Chaotic Time Series," *Phys. Rev. Lett.* **55**, 1–4 (1985).
190. W. M. SCHAFFER AND M. KOT, "Differential Systems in Ecology and Epidemiology," in *Chaos,* A. V. Holden, Ed. (Princeton University Press, Princeton, 1986).
191. H. G. SCHUSTER, *Deterministic Chaos* (VCH Weinheim, Germany, 1988).
192. I. SCHWARTZ, "Multiple Recurrent Outbreaks and Predictability in Seasonally Forced Nonlinear Epidemic Models," *J. Math. Biol.* **21**, 347–361 (1985).
193. L.-H. SHAN, P. HANSEN, C. K. GOERTZ AND R. A. SMITH, "Chaotic Appearance of the AE Index," *Geophys. Rev. Lett.* **18**, 147–150 (1991).

194. M. B. SHARIFI, K. P. GEORGAKAKOS AND I. RODRIGUEZ-ITURBE, "Evidence of Deterministic Chaos in the Pulse of Storm Rainfall, *J. Atmos. Sci.* **47**, 888–893 (1990).

195. R. SHAW, C. D. ANDERECK, L. A. REITH AND H. L. SWINNEY, "Superposition of Travelling Waves in the Circular Couette System," *Phys. Rev. Lett.* **48**, 1172–1175 (1982).

196. J. A. SHEINKMAN AND B. LEBARON, "Nonlinear Dynamics and Stock Returns," *J. Business* 331–337 (1989).

197. R. H. SIMOYI, A. WOLF AND H. L. SWINNEY, "One-dimensional Dynamics in a Multicomponent Chemical Reaction," *Phys. Rev. Lett.* **49**, 245–248 (1982).

198. L. A. SMITH, "Intrinsic Limits on Dimension Calculations," *Phys. Lett. A* **133**, 283–288 (1988).

199. G. STRANG, *Introduction to Applied Mathematics.* (Wellesley-Cambridge Press, Wellesley, Massachusetts, 1986).

200. G. SUGIHARA AND R. M. MAY, "Nonlinear Forecasting as a Way of Distinguishing Chaos from Measurement Error in Time Series," *Nature* **344**, 734–741 (1990).

201. H. L. SWINNEY AND J. P. GOLLUB, "The Transition to Turbulence," *Phys. Today* **31**, 41–47 (1978).

202. H. L. SWINNEY AND J. C. ROUX, "Chemical chaos," in *Nonequilibrium Dynamics in Chemical Systems,* C. Vidal, Ed. (Springer, New York, 1984).

203. F. TAKENS, *Dynamical Systems and Turbulence.* Lecture Notes in Mathematics, vol. 898, (Springer, New York, 1981).

204. J. TESTA, J. PEREZ AND C. JEFFRIES, "Evidence for Universal Chaotic Behavior of a Driven Nonlinear Oscillator," *Phys. Rev. Lett.* **48**, 714–717 (1982).

205. J. THEILER, "Efficient Algorithm for Estimating the Correlation Dimension from a Set of Discrete Points," *Phys. Rev. A* **36**, 4456–4462 (1987).

206. J. THEILER, "Some Comments on the Correlation Dimension of $1/f^a$ noise," *Phys. Lett. A.* **155**, 480–492 (1991).

207. B. TOWNSHEND, "Nonlinear Prediction of Speech Signals," in *Nonlinear Prediction and Modeling,* M. Casdagli and S. Eubank, Eds. (Addison-Wesley, Reading, Massachusetts, 1991).

208. A. A. TSONIS, "The Effect of Truncation and Round-off on Computer Simulated Chaotic Trajectories," *Comp. Math. Appl.* **21**, 93–94 (1991).

209. A. A. TSONIS AND J. B. ELSNER, "Fractal Characterization and Simulation of Lightning," *Beitr. Phys. Atmos.* **60**, 187–192 (1987).

210. A. A. TSONIS AND J. B. ELSNER, "The Weather Attractor Over Very Short Time Scales," *Nature* **33**, 545–547 (1988).

211. A. A. TSONIS AND J. B. ELSNER, "Chaos, Strange Attractors and Weather," *Bull. Am. Met. Soc.* **70**, 16–23 (1989).

212. A. A. TSONIS AND J. B. ELSNER, "Multiple Attractors, Fractal Basins and Long-Term Climate Dynamics, *Beitr. Phys. Atmos.* **63**, 171–176 (1990a).

213. A. A. TSONIS AND J. B. ELSNER, "Comments on Dimension Analysis of Climatic Data," *J. Climate* **3**, 1502–1505 (1990b).

214. A. A. TSONIS AND J. B. ELSNER, "Nonlinear Prediction as a Way of Distinguishing Chaos from Random Fractal Sequences," *Nature* (in press).

215. A. A. TSONIS, J. B. ELSNER AND K. P. GEORGAKAKOS, "Estimating the Dimension of Weather and Climate Attractors: What Do We Know about the Procedure and What Do the Results Mean?", preprint, University of Wisconsin–Milwaukee, 1992.

216. A. A. TSONIS, J. B. ELSNER AND P. A. TSONIS, "On the Dynamics of a Forced Reaction-

diffusion Model for Biological Pattern Formation," *Proc. Nat. Acad. Sci. USA* **80,** 4938–4942 (1989).

217. A. A. Tsonis and P. A. Tsonis, "Fractals: A New Look at Biological Shape and Patterning," *Presp. Biol. Medic.* **30,** 355–361 (1987).

218. J. S. Turner, J. C. Roux, W. D. McCormick and H. L. Swinney, "Alternating Periodic and Chaotic Regimens in a Chemical Reaction-Experiment and Theory," *Phys. lett. A* **85,** 9–12 (1981).

219. D. V. Vassiliadis, A. S. Sharma, T. E. Eastman and K. Papadopoulos, "Low-dimensional Chaos in Magnetospheric Activity from AE Time Series," *Geophys. Res. Lett.* **17,** 1841–1844 (1990).

220. C. Vidal, J. C. Roux and S. Bachelart, "Experimental Study of the Transition to Turbulence in the Belousov-Zhabotinsky Reaction," *N.Y. Acad. Sci.* **357,** 377–390 (1980); D. J. Wales, "Calculating the Rate of Loss of Information from Chaotic Time Series by Forecasting," *Nature* **350,** 485–488 (1991).

221. *Weatherwise* "Forecasting into Chaos," **42,** 202–207 (1990).

222. P. J. Werbos, "Back-Propagation Through Time: What It Does and How to Do It," *Proc. IEEE* **78,** 1550–1560 (1990).

223. N. Wiener, "Nonlinear Prediction and Dynamics," in *Proceedings of the Third Berkeley Symposium.* J. Neyman, Ed. (University of California Press, Berkeley, 1956).

224. N. Wiener, *Cybernetics* (MIT Press, Cambridge, Massachusetts, 1961).

225. A. Wolf, J. B. Swift, H. L. Swinney and J. A. Vastano, "Determining Lyapunov Exponents from a Time Series," *Phys. D* **16,** 285–317 (1985).

226. G. M. Zavlavsky, R. Z. Sagdeev, D. A. Usikov and A. A. Chernikov, *Weak Chaos and Quasi-regular Patterns* (Cambridge University Press, Cambridge, 1991).

227. W.-M. Zhang, J.-M. Yuan, D. H. Feng, Q. Pan and J. Tjon, "Quantum Fluctuations in Classical Chaos," *Phys. Rev. A* **42,** 3646–3649 (1990).

INDEX